S343 Block 7

Science: a third level course

Block 7 **BIOINORGANIC CHEMISTRY**

The Open University

Prepared by an Open University Course Team

S343 Course Team

Course team chairs	*Stuart Bennett, Elaine Moore, Lesley Smart*
Authors	Block 1 *David Johnson*
	Block 2 *Elaine Moore*
	Block 3 *Kiki Warr, with contributions from David Johnson*
	Block 4 *Stuart Bennett*
	Block 5 *Michael Mortimer*
	Block 6 *Ivan Parkin, with contributions from Dr M. Kilner, Professor K. Wade (University of Durham) and F. R. Hartley (Royal Military College of Science, Shrivenham)*
	Block 7 *Paul Walton (University of York), with contributions from Lesley Smart*
	Block 8 *Lesley Smart, with contributions from David Johnson, Kiki Warr and Elaine Moore*
	Block 9 *David Johnson*
Consultants	*Dr P. Baker* (University College of North Wales)
	Dr R. Murray (Trent Polytechnic)
Course managers	*Peter Fearnley*
	Wendy Selina
	Charlotte Sweeney
Editors	*Ian Nuttall*
	David Tillotson
BBC	*Andrew Crilly*
	David Jackson
	Jack Koumi
	Michael Peet
Graphic artists	*Steve Best, Janis Gilbert, Andrew Whitehead*
Graphic designers	*Josephine Cotter, Sarah Hofton, Tag Taylor*
Assistance was also received from the following people:	*George Loveday* (Staff Tutor)
	Joan Mason
	Jane Nelson (Staff Tutor)
Course assessor	*Professor J. F. Nixon* (University of Sussex)

The Open University, Walton Hall, Milton Keynes, MK7 6AA

First published 1989; second edition 1997

Reprinted 2001

Copyright © 1989, 1997 The Open University

All rights reserved. No part of this publication may be reproduced, stored in a retrieval system or transmitted in any form or by any means, without written permission from the publisher or a licence from the Copyright Licensing Agency Limited. Details of such licences (for reprographic reproduction) may be obtained from the Copyright Licensing Agency Limited of 90 Tottenham Court Road, London W1P 0LP.

Edited, designed and typeset by the Open University.

Printed in the United Kingdom by Henry Ling Limited, at the Dorset Press, Dorchester, DT1 1HD.

ISBN 07492 8167 7

This publication forms part of an Open University course S343. [The complete list of texts which make up this course can be found at the back]. Details of this and other Open University courses can be obtained from the Call Centre, PO Box 724, The Open University, Milton Keynes MK7 6ZS, United Kingdom: tel. +44 (0)1908 653231, e-mail ces-gen@open.ac.uk.

Alternatively, you may visit the Open University website at http://www.open.ac.uk where you can learn more about the wide range of courses and packs offered at all levels by the Open University.

To purchase this publication or other components of Open University courses, contact Open University Worldwide Ltd, The Berrill Building, Walton Hall, Milton Keynes MK7 6AA, United Kingdom: tel. +44 (0)1908 858785; fax +44 (0)1908 858787; e-mail ouwenq@open.ac.uk; website http://www.ouw.co.uk

STUDY GUIDE FOR BLOCK 7

1 INTRODUCTION — BIOINORGANIC CHEMISTRY

2 ELEMENTS AND LIVING SYSTEMS
- 2.1 Fluorine
- 2.2 Elements in living systems
- 2.3 Biomolecule study

3 INORGANIC ELEMENTS AND BIOCHEMICAL MOLECULES
- 3.1 Amino acids
- 3.2 Porphyrins
- 3.3 Nucleic acids as ligands
- 3.4 Summary of Section 3

4 BIOMOLECULAR STUDY: PROBLEMS AND SOLUTIONS
- 4.1 Single-crystal X-ray diffraction (XRD)
- 4.2 Extended X-ray absorption fine structure (EXAFS)
- 4.3 Resonance Raman spectroscopy
- 4.4 Summary of Section 4

5 OXYGEN TRANSPORT IN LIVING SYSTEMS
- 5.1 Introduction
- 5.2 The chemistry of oxygen
- 5.3 O_2 transport
  - 5.3.1 Introduction
  - 5.3.2 Haemoglobin and myoglobin
  - 5.3.3 Myoglobin structure
  - 5.3.4 Haemoglobin
  - 5.3.5 Haemerythrin and haemocyanin — two different O_2-carriers
- 5.4 Summary of Section 5

6 O_2-ACTIVATING AND H_2O_2-ACTIVATING ENZYMES
- 6.1 O_2 in aerobic respiration — cytochrome c oxidase
- 6.2 Cytochrome P450
- 6.3 Peroxidases, catalases and Cu–Zn superoxide dismutase
- 6.4 Summary of Section 6

7 IRON TRANSPORT AND STORAGE
- 7.1 Principles of iron chemistry: the problems of iron uptake
  - 7.1.1 Summary of iron chemistry
- 7.2 Iron uptake by organisms
  - 7.2.1 Removal of iron
  - 7.2.2 Summary of Section 7.2
- 7.3 Iron transport and storage
  - 7.3.1 Iron transport
  - 7.3.2 Iron storage
- 7.4 Summary of Section 7

8 ZINC: A CASE STUDY
- 8.1 Zinc chemistry
- 8.2 Zinc(II): biology's Lewis acid
  - 8.2.1 Carbonic anhydrase
  - 8.2.2 Liver alcohol dehydrogenase
- 8.3 Zinc(II) in a structural role
- 8.4 Summary of Section 8

9 THE FUTURE OF BIOINORGANIC CHEMISTRY

GLOSSARY OF BIOINORGANIC TERMS	90
OBJECTIVES FOR BLOCK 7	96
SAQ ANSWERS AND COMMENTS	98
ACKNOWLEDGEMENTS	106

STUDY GUIDE FOR BLOCK 7

Block 7 is equivalent to two weeks of study time, and contains the bioinorganic component of S343. You should aim to reach the end of Section 5 by the end of the first week.

Since the subject matter discussed contains many biological terms that you may not have come across since the Science Foundation Course, we have added a Glossary. The first appearance of each glossary term is marked by a superscript 'G'.

There is no video programme to accompany this Block. The video programme on VC2 for Block 7 is on nitrogen fixation, and refers to an earlier version of the Block.

Bioinorganic chemistry is the study of inorganic elements in living systems. Many of the topics we shall study in this rapidly expanding area of modern chemistry are very complex, and yet they can be understood in terms of basic inorganic chemistry. As such, we shall draw heavily on your previous knowledge of both inorganic chemistry — especially about the transition and main-Group metals — and of biochemistry. We shall build on the major theories of chemistry that you have studied in the first part of the Course. You may find it helpful to have Blocks 1, 2, 3 and 4 of S343 and Units 22 and 24 of S102 to hand.

Bioinorganic chemistry is a large area of chemistry and it is impossible to cover the whole subject in a single Block; instead we shall study particular examples of the subject. You are not expected to learn all the individual reactions and detailed structures in the Block. You may find it helpful to use your model kit to make models of some of the molecules discussed.

1 INTRODUCTION — BIOINORGANIC CHEMISTRY

What is bioinorganic chemistry? The word 'bioinorganic', which is a composite of biology and inorganic, is used to describe the study of the occurrence and properties of inorganic elements (that is, elements other than carbon, hydrogen and oxygen) in living systems. At first sight this may appear to be a rather narrow field of study, since nearly all living matter is composed of organic compounds. However, it is a fact that without certain inorganic elements *no* organism could exist.

Table 1 shows the **recommended daily allowances (RDA)** of certain elements for humans. (A similar list can be found on the back of any vitamins and 'minerals' bottle.) A deficiency or excess in our diets of any one of these elements leads to health problems. From this list it is clear that inorganic elements are an integral and essential part of our biochemical systems. By studying these elements in a biological context, not only can we learn what roles they play in biology, but we can also provide a foundation for understanding, and eventually treating, many of the health problems listed in Table 1. Furthermore, as you will see later in this Block, the properties of molecules containing inorganic elements that are found in biochemical systems are often unexpected and completely different from the inorganic compounds you have met before.

Table 1 Recommended daily allowance (RDA) of certain elements* for a human adult, and the effect of deficiency in the diet.

Inorganic element	RDA/mg	Effect of deficiency
sodium	1 100–3 300	†
chlorine	3 200	†
potassium	2 000–5 500	†
magnesium	300–400	muscle cramps
calcium	800–1 200	retarded skeletal growth
fluorine	1.5–4.0	dental decay
iodine	0.15	thyroid disorders, retarded metabolism
chromium	0.05–0.25	diabetes symptoms
molybdenum	0.075–0.25	poor cell growth
manganese	2–5	infertility
iron	10–20	anaemia^G, immune system disorders
cobalt (as vitamin B_{12})	2	pernicious anaemia
copper	1.5–3	liver disorders, artery weakness
zinc	15	skin damage, stunted growth

* This is not an exhaustive list.

† Sodium, chlorine and potassium deficiency is rare, and acute cases only tend to occur with severe dehydration.

SAQ 1 As you work through Block 7, draw up a table of the metalloproteins^G you encounter, indicating the metal(s) that is(are) found there and the function of the protein.

2 ELEMENTS AND LIVING SYSTEMS

Figure 1 shows a 'biological Periodic Table'. The elements that are known to occur in biochemical systems (in both plants and animals) are shown in green. These elements can be further classified into either *bulk* or *trace*. **Bulk** means that the element makes up a significant percentage by mass of most organisms (say > 0.1%); **trace** means that the element is present in most organisms in very small quantities.

Group	I	II										III	IV	V	VI	VII	0
1st Period	**H**																He
2nd Period	Li	Be										*B*	*C*	*N*	*O*	*F*	Ne
3rd Period	**Na**	**Mg**										Al	*Si*	*P*	*S*	*Cl*	Ar
4th Period	**K**	**Ca**	Sc	Ti	*V*	*Cr*	*Mn*	*Fe*	*Co*	*Ni*	*Cu*	*Zn*	Ga	Ge	*As*	*Se*	*Br*	Kr
5th Period	Rb	Sr	Y	Zr	Nb	*Mo*	Tc	Ru	Rh	Pd	Ag	Cd	In	*Sn*	Sb	Te	*I*	Xe
6th Period	Cs	Ba	La	Hf	Ta	W	Re	Os	Ir	Pt	Au	Hg	Tl	Pb	Bi	Po	At	Rn
7th Period	Fr	Ra	Ac														

Figure 1 Biological Periodic Table of the elements. Elements that occur naturally in biological systems are printed in green. Bulk elements are shown in bold and trace elements are shown in italics.

2.1 Fluorine

Fluorine is a very good example of a trace element that has a fascinating bioinorganic chemistry. (For the purposes of this study, fluorine is always in its fluoride form, F^-, and the term *fluoride* will be used for the remainder of the Block.) Fluoride is familiar to us as one of the active ingredients in toothpaste. Indeed, the major source of fluoride in Western human diets is from fluoride toothpaste, although some countries deliberately fluorinate drinking water. Fluoride is very effective in the prevention of dental caries (tooth decay) because it reacts with hydroxyapatite ($Ca_{10}(PO_4)_6(OH)_2$) — a major component of tooth enamel — to give a tenacious coating of fluoroapatite ($Ca_{10}(PO_4)_6F_2$). The chemistry of this process is:

$$Ca_{10}(PO_4)_6(OH)_2 \xrightarrow{F^-} Ca_{10}(PO_4)_6F_2 \qquad 1$$
$$\text{hydroxyapatite} \qquad\qquad \text{fluoroapatite}$$
$$pK_{sp} \approx 13 \qquad\qquad pK_{sp} \approx 14$$

Comparing solubility products* shows that fluoroapatite is roughly ten times less soluble than hydroxyapatite. In humans, fluoride is found almost entirely in the teeth and bones.

Fluoride serves as a good example of the delicate balance between beneficial and toxic levels of intake. Too little fluoride can lead to tooth decay; too much is very toxic. Despite the obvious benefits of some fluoride intake, there is considerable debate about the appropriate level of intake in humans. Many of the toxic effects of high levels of fluoride intake, such as poor healing of wounds and brittle bones, are thought to stem from fluoride's ability to coordinate to metals, giving sparingly soluble metal fluorides. This can be seen in Table 2, which lists the solubility products of Group II metal fluorides. The low values of K_{sp} show that insoluble fluoride salts of magnesium and calcium will form even at low concentrations of fluoride. Of greatest concern is the *in vivo* formation of calcium fluoride, CaF_2. Calcium, another bioinorganic element, is an essential part of many biochemical processes, and is a major constituent of bone. Any depletion of soluble calcium ion concentrations *in vivo* (by precipitation as CaF_2) can lead to a poor state of health.

Table 2 Solubility products of Group II fluorides

Metal salt	Solubility product, K_{sp}/mol^3 l^{-3}
MgF_2	7.42×10^{-11}
CaF_2	1.46×10^{-10}
SrF_2	4.33×10^{-9}
BaF_2	1.84×10^{-7}

SLC 1

SLC 2

For example, the interference of fluoride with calcium biochemistry can be seen in the highly toxic nature of *hydrofluoric acid*, HF. HF is a weak acid, $pK_a = 3.17$ {cf. pK_a (HCl) ≈ −7; pK_a (ethanoic acid) = 4.76 and pK_a (HNO$_3$) ≈ −1.4, where pK_a = $\log_{10}$(*acid dissociation constant*^{G})}. Despite its comparatively weak acidity, hydrofluoric

---

* **Solubility product** is a measure of the solubility of a compound in water at pH 7 (and usually 298 K). For a sparingly soluble salt, M_aX_b, in contact with its saturated aqueous solution,

$$M_aX_b(s) = aM^{b+}(aq) + bX^{a-}(aq) \qquad 2$$

the solubility product is given by

$$K_{sp} = [M^{b+}(aq)]^a[X^{a-}(aq)]^b \qquad 3$$

The value of the solubility product can be expressed logarithmically in an analogous fashion to pK_a:

$$pK_{sp} = -\log_{10}(K_{sp})$$

Accordingly, a high value of pK_{sp} or a low value of K_{sp} indicates that the compound is only sparingly soluble. (See also, Block 3 SAQ 8.)

acid causes very serious skin burns that are slow to heal. The slow healing is almost certainly due to the formation of CaF_2 precipitates, which depletes the amount of soluble calcium ions in the vicinity of the burn. Calcium is an essential part of some of the proteins and enzymes[G] that are involved in the complex biochemistry of healing. Reducing the amount of available calcium ions reduces the efficiency of these proteins and enzymes.

There is an optimal intake level of fluoride, just as there is for all bioinorganic elements, in the form shown schematically in Figure 2. The overall shape of the curve in Figure 2 is common to all the bioinorganic elements shown in Figure 1. The position and broadness of the curve differ widely from element to element and from organism to organism. For example, mercury is toxic to humans in very low concentrations. (The term 'mad-as-a-hatter' originally referred to milliners who used $Hg(NO_3)_2$ to polish hat felts. Long-term, low-level exposure to mercury is now known to lead to dementia.) Despite the high toxicity of mercury to humans, some microorganisms have a much greater tolerance of high mercury concentrations. Unlike humans, these remarkable microorganisms are able to dispose of ingested mercury by biochemically converting it to dimethylmercury, $[Hg(CH_3)_2]$. Dimethylmercury is relatively volatile (b.t. 93 °C), and simply evaporates from the microorganism. As a result, these microorganisms are able to exist in mercury-contaminated areas. The mercury state-of-health curve for the microorganisms is much broader than the same curve for humans.

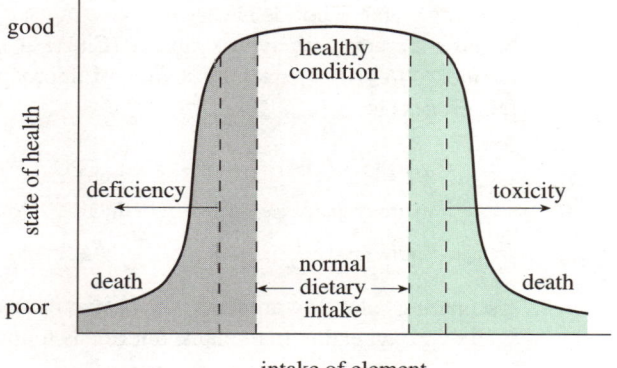

Figure 2 Schematic diagram of 'state of health' versus intake concentration of a particular element.

2.2 Elements in living systems

Other elements, such as iron, are required by almost all organisms. Iron has an essential role to play in many enzymes and proteins. In mammals, including humans, iron is an integral part of the blood; a deficiency leads to anaemia, a condition that affects many people. Accordingly, the bioinorganic chemistry of iron has been studied extensively, with most studies concentrating on the blood protein haemoglobin[G]. In Section 5, this protein along with the closely related protein myoglobin[G], will be discussed in detail. We shall see the essential role that iron plays in these proteins.

Almost all organisms require iron to live, and the availability of iron may be a limiting factor to an organism's survival. The iron content of the Earth's continental crust is actually relatively high (7.1%), but most of it exists as the highly insoluble — and, therefore, unassimilable — compounds, hydrated iron oxide (haematite), $Fe_2O_3.nH_2O$, iron hydroxide, $Fe(OH)_3$, magnetite, Fe_3O_4, or siderite, $FeCO_3$. Efficient 'capture' of iron and other inorganic elements is essential if an organism is to survive. The ability of organisms to capture selectively inorganic ions is exemplified by the concentration factors in human blood plasma shown in Table 3. From the Table, it can be seen that the concentrations of ions in human blood plasma are markedly different from the concentration of the same ion in seawater. These differences in concentration are an indirect measure of the active uptake of inorganic ions by humans, and organisms in general*.

---

* The basis for correlating seawater concentrations with human plasma concentrations stems from the assumption that all living species derive from the primordial organisms that once lived in the oceans. The concentration of inorganic ions in human plasma is, therefore (and rather crudely), assumed to be the same as the concentration of inorganic ions within these primordial organisms.

Table 3 Comparison of approximate concentrations of metal ions in seawater with human plasma; 'concentration factor' is [human plasma]/[seawater], and gives an indication as to how the particular element is 'concentrated' by humans

Element	Seawater conc./ mol l^{-1} × 10^{-8}	Human plasma conc./ mol l^{-1} × 10^{-8}	Concentration factor
sodium	4.6 × 10^7	2 × 10^5	c. 4 × 10^{-3}
magnesium	5.3 × 10^6	9 × 10^4	c. 0.02
potassium	9.7 × 10^5	2 × 10^5	c. 0.2
calcium	1.0 × 10^6	1 × 10^6	c. 1
vanadium	4.0	17.7	c. 4
chromium	0.4	5.0	c. 14
manganese	0.7	10.9	c. 15
iron	0.005–2*	2 230	1 100–450 000
cobalt	0.7	0.002 5	3.6 × 10^{-3}
nickel	0.5	10.4	c. 21
copper	1.0	1 650	1 650
zinc	8.0	1 720	215
molybdenum	10.0	1 000	1 000

* Dependent on pH, which may vary locally from its average value of 8.2, and on the anion content of seawater.

For some elements, such as iron and copper, the difference in concentrations is very high. Certainly, in humans at least, these ions are actively taken up in an energy-consuming (endothermic) process. On the other hand, cobalt ions have a relatively high concentration in seawater but a very low concentration in plasma; in this case it appears as if cobalt ions are prevented from entering the body's cells. The biochemical processes that are involved in metal uptake are complicated. Nevertheless, an understanding of metal-uptake systems is important in the successful treatment of the health disorders shown in Table 1. The uptake mechanisms are an active area of research, and, in Section 7, biochemical metal-uptake systems will be described. In the same Section, the human body's remarkable ability to store and transport iron will be discussed.

In Sections 6 and 8, the question, 'How are metals utilised within biochemical systems?' will be addressed. Section 6 will concentrate on the activation of oxygen by metal-containing proteins. Many of the reactions that are catalysed by metal-containing proteins are remarkably complex, and still cannot be mimicked by modern synthetic chemistry. Section 8 will examine in detail some zinc-containing proteins. Zinc(II) is used as nature's Lewis acid[G] in a wide range of enzymes, where it is often an important part of the enzyme's catalytic mechanism. Zinc(II) is also found in other proteins, including the important zinc-finger proteins[G], which recognise DNA (deoxyribonucleic acid)[G], where the zinc(II) ion is essential in maintaining the protein's three-dimensional structure. On genetic evidence, it has been suggested that there may be up to 200 zinc-finger proteins waiting to be characterised.

2.3 Biomolecule study

Our current knowledge of bioinorganic chemistry comes from direct studies of inorganic elements in biological systems. For these studies a range of physical techniques is available to the bioinorganic chemist. Some of these techniques will be discussed in Section 4, including single-crystal X-ray diffraction[G], extended X-ray absorption fine structure, EXAFS[G] and, in particular, resonance Raman spectroscopy[G]. Of these techniques, single-crystal X-ray diffraction is the most revealing, in that it gives an accurate picture of the three-dimensional structure of a protein. Structural information about proteins is essential for a fuller understanding of the way proteins and enzymes function.

Bioinorganic chemistry is a rapidly expanding area. What is currently known about inorganic elements in living systems is only the tip of the iceberg, and much is waiting to be discovered. To illustrate the rapid expansion of this area and as an indication of the remarkable use of metals in living systems, the structure of a recently discovered magnesium-containing protein (called 'light-harvesting complex II') is shown below. (The structure was determined using single-crystal X-ray diffraction.) The magnesium ion is an essential part of the protein; it is likely that if any other metal ion replaced magnesium in this protein, then the protein would not function properly. The protein is found in photosynthetic bacteria, which can produce energy via photosynthesis. The protein's role is to 'collect' light energy for use in the photosynthetic biochemical reaction. It is called a light-harvesting protein. Its structure was first published in 1995 and is truly astonishing. The details of the structure are shown in Figures 3 and 4 (and also in Plate 1, inside the front cover). Figure 3 shows a magnesium-containing bacteriochlorophyll-a (bch-a) molecule, which is capable of absorbing visible light (the plant pigment chlorophyll[G] also contains magnesium). Figure 4 depicts the arrangement of the bch-a molecules within the light-harvesting protein. Part of the structure consists of eighteen bch-a molecules stacked in a ring, where each bch-a molecule overlaps two others. It has been proposed that the ring acts as a storage ring for light energy. The energy is temporarily 'stored' here on a nanosecond time-scale until it is required to provide energy for a photosynthetic reaction. The structure is unprecedented, and is an outstanding example of the occurrence and use of metal complexes in living systems.

O
N
N
Mg
N
N
O
O
O
O
O—CH3
R
CH3
CH3
CH3
CH3
R =
H3C

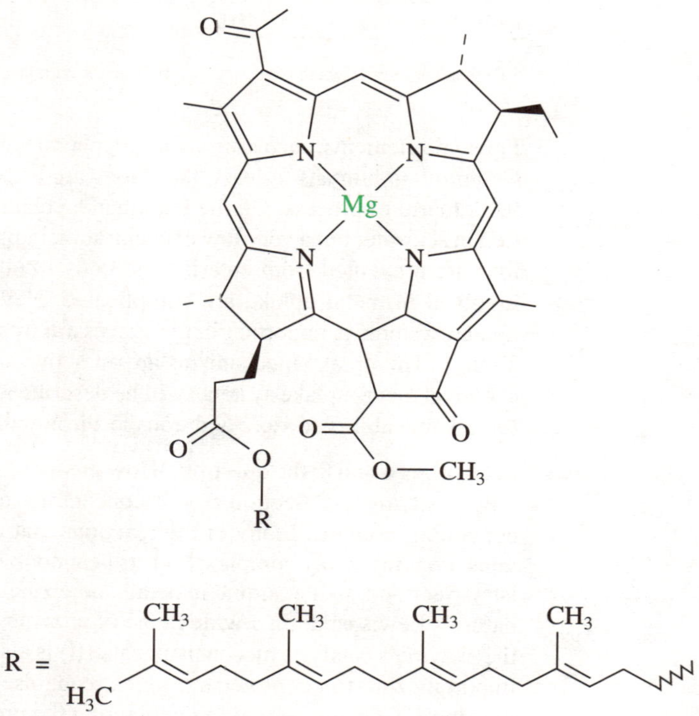

Figure 3 Structure of bacteriochlorophyll-a. In this and subsequent structures the metal atom(s) is(are) printed in green.

α-helix viewed
down axis of helix
bacteriochlorophyll-a
viewed edge on

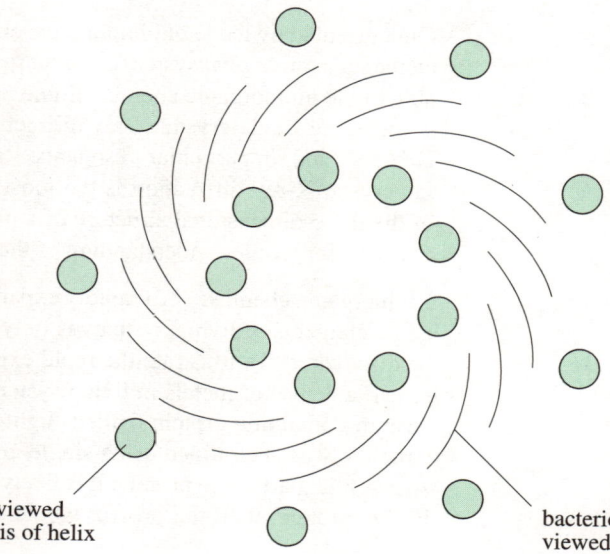

Figure 4 Schematic representation of part of the light-harvesting protein structure.

3 INORGANIC ELEMENTS AND BIOCHEMICAL MOLECULES

As we have already seen in the previous Section, certain inorganic elements are essential for life. Table 4 shows the amounts and mole fractions (expressed relative to the important bioinorganic element iron) of some bioinorganic elements in a 70 kg human. The table shows three clear classes of elements. The bulk elements, calcium, phosphorus, sulphur, sodium, potassium, chlorine and magnesium, form one class. These elements (mostly in ionic form) occur widely throughout the body. Calcium is particularly abundant due to its presence in bone; it is also found in a variety of proteins and enzymes. A second class contains the trace elements iron, silicon, fluorine and zinc, which are found in small quantities. Both iron and zinc ions are found in blood proteins, and are essential for absorbing and discharging both oxygen and carbon dioxide. Silicon and fluorine occur in bone. A third class contains elements that are found in *very* small amounts.

Table 4 Masses and mole fractions (relative to iron) of bioinorganic elements in a typical 70 kg human

Element	Mass/g	Mole fraction (relative to Fe)*
calcium	1 050	348
phosphorus	700	301
sulphur	175	72.8
sodium	105	60.7
potassium	140	47.6
chlorine	105	39.5
magnesium	35	19.1
iron	4.2	1.00
silicon	1.4	0.66
fluorine	0.8	0.56
zinc	2.3	0.47
bromine	0.2	0.033
copper	0.11	0.023
manganese	0.02	0.005
vanadium	0.02	0.005
lithium	0.002	0.004
selenium	0.02	0.003
chromium	0.005	0.001
cobalt	0.003	0.000 7

* The number of moles of a particular element A, n_A, divided by the number of moles of iron, n_{Fe}, in an average 70 kg human, n_A/n_{Fe}.

Remarkably, there is very little variation in the mole fractions from one person to another. Indeed, any significant variation leads to disease. This prompts the question: What controls the uptake, storage and use of bioinorganic elements within living systems? Clearly, within an organism, specific molecules must associate with particular bioinorganic elements. This Section describes some of the biochemical molecules which are known to interact directly with bioinorganic elements. At the same time, the underlying chemical concepts will be discussed. Later Sections will describe specific cases in more detail.

In Block 4 we saw how ligands interact with metals. The same basic chemical principles of coordination chemistry apply in bioinorganic chemistry, in so far as metals in living systems are always associated with ligands in coordination complexes. Most of the 'biological ligands' are familiar biochemical molecules such as amino acids[G], saccharides[G], nucleotides[G] (purines and pyrimidines), vitamins[G] and hormones[G]. Biochemical systems also make use of a range of other ligands, including the porphyrin[G] group that we met in Block 4.

3.1 Amino acids

Of all the biochemical ligands the most common are amino acids. The most important amino acid side-chains in bioinorganic chemistry are shown in Figure 5*. The characteristic feature of these amino acid side-chains is an ability to coordinate to a metal ion via a lone pair of electrons on an electronegative atom.

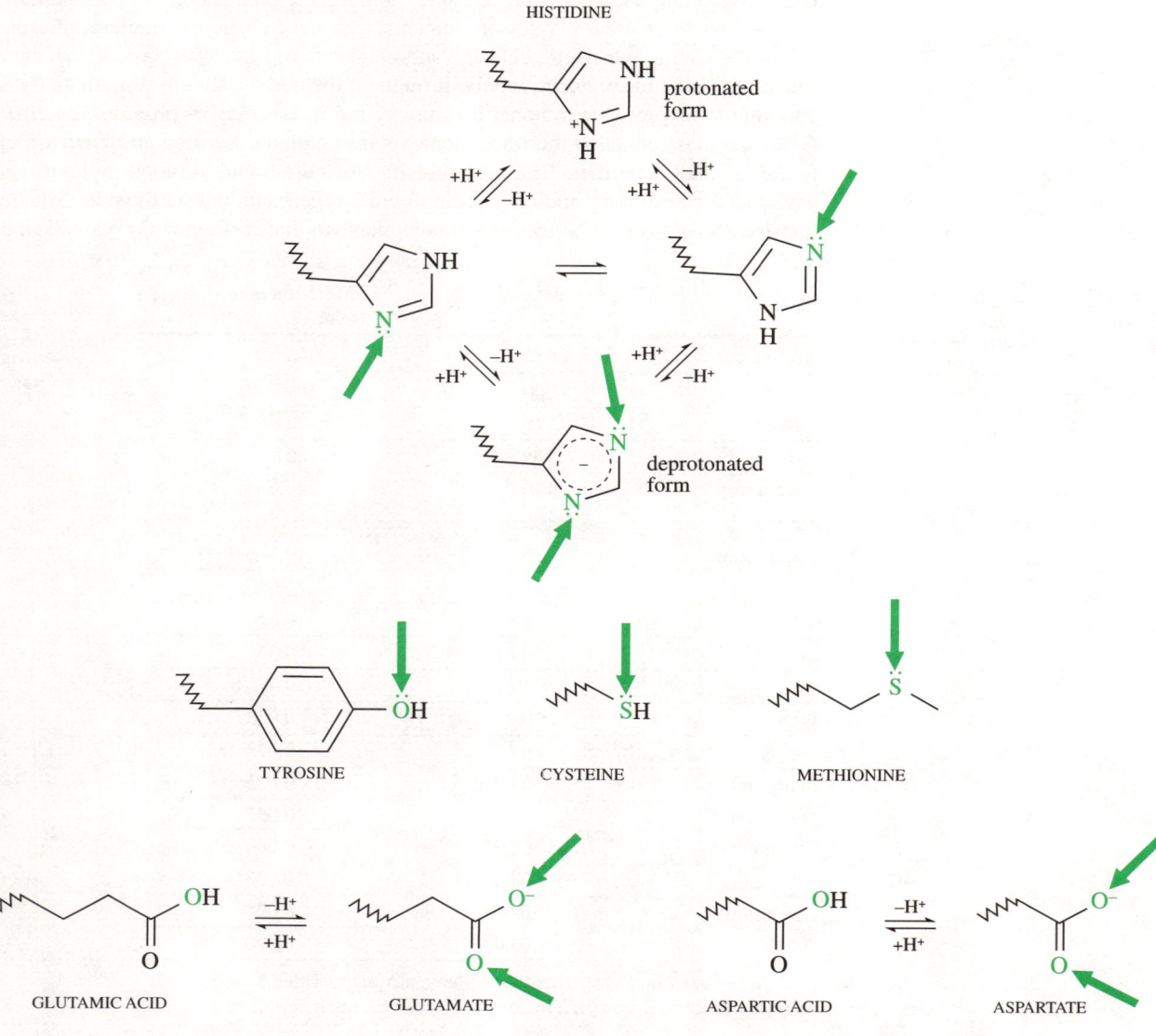

Figure 5 The amino acid side-chains most commonly found to coordinate to metal ions in biochemical molecules. The green arrows denote the position of coordination to a metal ion.

☐ Through which atom would you expect histidine to bond to a metal ion?

■ Histidine has a five-membered ring as part of its side-chain. In its neutral form, one of the nitrogen atoms in the ring has a lone pair that can form a coordination bond to a metal ion.

Note that the neutral histidine can be protonated; in its protonated form it is unable to coordinate to a metal ion using a nitrogen lone pair, and it will not form stable complexes with metal ions. The side-chain of neutral histidine can also be deprotonated to give a histidate anion.

☐ How does deprotonation to form the histidate anion affect its coordination properties?

■ The side-chain can now coordinate to two separate metal ions simultaneously.

In a similar fashion, aspartate and glutamate are the deprotonated forms of aspartic acid and glutamic acid. The deprotonated (carboxylate) forms can act as ligands in stable complexes with many metal ions.

* The side-chains of all twenty naturally occurring α-amino acids are given in the Glossary as Table 11.

☐ Would you also expect that protonated (acid forms) glutamic acid and aspartic acid form complexes with metal ions? If so, how?

■ Yes. They can coordinate using the lone pairs on the carbonyl oxygen. These complexes are not as stable as the carboxylate–metal ion complexes. The reason for this is that there is not the same extent of coulombic attraction between the acid form and the metal ion as with the carboxylate form and metal ion.

Therefore, the protonation state of the amino acid side-chain is an important factor in determining whether the side-chain will form stable complexes with metal ions. A measure of the extent of protonation (at a given pH and temperature in aqueous solution) of a particular chemical group is given by the acid dissociation constant, K_a, and the pK_a of the acid form. The approximate pK_as of the protonated forms of bioinorganically important amino acid side-chains are given in Table 5.

Table 5 Amino acid side-chains with pK_a values (together with the classification of the ligand as a hard or soft acid[G]; see later); the acidic proton is shown in green

Amino acid side-chain		pK_a	Hard/soft
(imidazole with NH+)	protonated histidine	6.5	—*
(imidazole)	histidine	14	hard/borderline
(–COOH)	aspartic acid	4.5	hard (carboxylate form)
(–CH₂COOH)	glutamic acid	4.5	hard (carboxylate form)
(–C₆H₄–OH)	tyrosine	10	hard (deprotonated form)
(–SH)	cysteine	8.5	soft
(–S–CH₃)	methionine	—	soft

* The side-chain of protonated histidine cannot act as a Lewis base[G] because it has no lone pair of electrons available.

We use K_a and pK_a to determine the extent of protonation. If we denote the protonated (acidic) form of the side-chain as AH, and the deprotonated (basic) form as B, then we can write an equation for the acid/base equilibrium*

$$AH(aq) = B(aq) + H^+(aq) \qquad 4$$

☐ Write an expression for K_a for this reaction.

■ $$K_a = \frac{[B(aq)][H^+(aq)]}{[AH(aq)]}$$

By rearranging this expression, we can see that the ratio of the concentration of base to the concentration of acid form, $[B(aq)]/[AH(aq)]$, is given by $K_a/[H^+(aq)]$.

So, to determine the ratio of base to acid in any aqueous solution, we need to know both K_a and the acidity, which is given by the pH. The pH of most biochemical systems tends to be around pH 7.

☐ How do you calculate $[H^+(aq)]$ at pH 7?

■ $pH = -\log_{10}[H^+(aq)] = 7$, and thus $[H^+(aq)] = 10^{-7}$ mol l^{-1}.

Let's take the deprotonation reaction of neutral histidine as a first example. We see from Table 5 that the pK_a is 14. Notice that the pK_a value is *higher* than the pH of most biochemical systems (c. pH 7). We shall come back to this later.

☐ What is the value of K_a for the deprotonation of histidine?

■ $pK_a = -\log_{10} K_a$, so $K_a = 10^{-14}$ mol l^{-1}.

The base/acid ratio for histidine at pH 7 is therefore $10^{-14}/10^{-7}$ or 10^{-7}, and so very little histidine is deprotonated at pH 7.

Now let's take aspartic acid as an example (glutamic acid is similar). In this case the pK_a value is 4.5. Notice that pK_a is now lower than the pH of most biochemical systems (c. pH 7).

☐ What is the ratio of the concentrations of aspartate to aspartic acid?

■ The ratio is given by $10^{-4.5}/10^{-7} = 10^{2.5}$.

In this case we see that the base form, aspartate, dominates by a factor of about 300.

These two examples demonstrate a general point: if we know the pH of the system — and we can assume pH 7 for biochemical systems — then the pK_a value for the acid–base equilibrium reaction of the side-chain immediately tells us which form of the side-chain dominates. So, if the pK_a is greater than 7, the protonated form dominates, and if the pK_a is less than 7, the deprotonated form dominates. Thus, the Table shows that at pH 7, *in the absence of metal ions*, aspartic acid and glutamic acid exist *mostly* as aspartate and glutamate.

---

* Strictly speaking, the acid dissociation constant is defined in terms of the following equilibrium

$$AH(aq) + H_2O(l) = B(aq) + H_3O^+(aq)$$

which takes account of the fact that free protons, H$^+$, do not exist in aqueous solution. The ensuing analysis is based on the simpler equation 4, but comes to the same conclusions as if we had used the more accurate equation above.

Another question to ask is, by how much is the concentration of one form greater than the other? At pH 7, aspartate and glutamate as we have seen are $10^{2.5}$ times (or × 316) more abundant than the protonated forms. But what about the protonated forms of histidine compared with neutral histidine (this system has a pK_a of 6.5)?

☐ Work out the ratio of His(aq) to HisH⁺ at pH 7. (Assume that ionisation to the histidate anion is small enough to be ignored at pH 7.)

■ If we write the equilibrium

$$\text{HisH}^+(\text{aq}) \rightleftharpoons \text{His}(\text{aq}) + \text{H}^+(\text{aq}) \qquad 5$$

$$K_a = \frac{[\text{His}(\text{aq})][\text{H}^+(\text{aq})]}{[\text{HisH}^+(\text{aq})]} \qquad 6$$

and therefore,

$$\frac{[\text{His}(\text{aq})]}{[\text{HisH}^+(\text{aq})]} = \frac{K_a}{[\text{H}^+(\text{aq})]} \qquad 7$$

$pK_a = 6.5$, so $K_a = 10^{-6.5}$, giving

$$\frac{[\text{His}(\text{aq})]}{[\text{HisH}^+(\text{aq})]} = \frac{10^{-6.5}}{10^{-7}} = 10^{0.5} \qquad 8$$

$$= 3.16$$

The neutral form of histidine dominates, but now only by a factor of about three times; in other words the solution contains about 75 per cent histidine and 25 per cent of the protonated histidine. So histidine exists mostly in its neutral form at pH 7. Therefore, we can expect free histidine to be neutral, and aspartic and glutamic acids to be deprotonated at pH 7. The pK_as of cysteine and tyrosine (8.5 and 10, respectively) indicate that these amino acid side-chains will be protonated at pH 7.

However, we also need to remember that the pK_as listed in Table 5 are measured *in the absence of metal ions*, which interact strongly with the amino acids. In neutral aqueous solution, metal ions may displace the acidic proton of an amino acid side-chain, even in amino acids with pK_as higher than 7. In these cases, we have a different equilibrium reaction, in which a side-chain such as cysteinyl, which we can denote as RSH, reacts with a metal ion M^{n+}. So, it is possible to observe deprotonated cysteine coordinated to a metal ion as follows:

$$\text{RSH} + M^{n+} = [\text{RSH}-M]^{n+} = [\text{RS}-M]^{(n-1)+} + \text{H}^+ \qquad 9$$

The tyrosine side-chain, with a relatively high pK_a, is less likely to have its proton displaced by a metal ion; even so, tyrosyl side-chains can coordinate to a metal in both protonated and deprotonated forms. It is also possible to observe a deprotonated histidyl side-chain coordinated to two metal ions; that is, the proton has been displaced by a second metal ion.

The amide link, R—CO—NH—R′, in a polypeptide^G chain does not often coordinate to metals in bioinorganic chemistry and can be ignored as a coordinating group for our purposes. One of the reasons why the amide group is relatively poor at coordinating to metal ions can be seen from structures **1** and **2**, which are the two important resonance forms of the amide group:

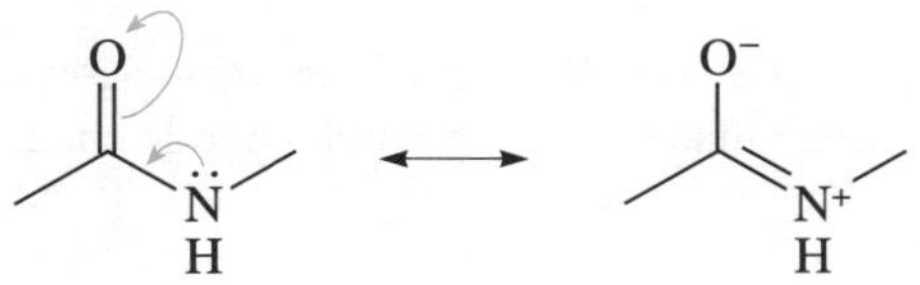

An important resonance form of the amide group is one in which the orbital containing the nitrogen lone pair of electrons forms part of a π molecular orbital between the nitrogen, carbon and oxygen atoms. In this form (**2**) we can consider the nitrogen to be sp^2 hybridised. As such, the nitrogen lone pair electrons are delocalised over the amide group, and are 'unavailable' for coordination to a metal ion. Even so, the oxygen atom would still be available for coordination; in fact, amide oxygens are only rarely found to be coordinating atoms in bioinorganic chemistry. Furthermore, the pK_a of the amide group is approximately 17.

☐ What is the consequence of the high pK_a of the amide group?

■ Ionisation of the proton requires a lot of energy, and so it is unlikely that a metal ion could displace the proton to form a deprotonated amide–metal ion complex.

SAQ 2 What other amino acids, not in Figure 5 and Table 5, could coordinate to metals? How might they coordinate? (See Table 11, p. 91 for the structures of the amino acid side-chains.)

As well as listing pK_a values, Table 5 also classifies amino acid side-chains according to the hard/soft acid–base theory introduced in Block 4. We can see that amino acids cover a wide range of hard to soft ligands. For instance, cysteine, which coordinates through sulphur, can be classified as a soft ligand.

☐ What kind of metal would you expect cysteine to coordinate to?

■ Soft ligands tend to coordinate to soft metal ions, including second- and third-row transition metal ions, particularly those in low oxidation states.

Aspartate, on the other hand, which coordinates through oxygen, is a hard ligand, and will coordinate effectively to hard metal ions such as those of the lighter alkaline earths.

This range of hardness is important when it comes to using a polypeptide to coordinate to a particular metal. The hardness of the coordinating amino acid side-chains will determine, to some extent, which metal is preferentially coordinated by that amino acid. Bone is a good example of this. Bone, as we have already seen, contains calcium ions.

☐ Is Ca^{2+} classified as a hard or soft acid?

■ Hard acid. Hard acids include the lighter alkali metals and alkaline earths, and the first-row transition metals in high oxidation states. Hard acids and bases have low polarisability, whereas the soft acids and bases tend to be more polarisable. A comprehensive list is given in Table 2 of Block 4.

Let's ask the question: how is broken bone repaired by the body? Or more specifically from a chemical point of view: how are calcium ions selectively deposited onto the area of the break? Part[*] of the answer to this latter question can be found by examining the chemical make-up of bone in Table 6.

Table 6 Chemical content of bone

Material	Mass/per cent
fibrous protein (collagen)	≈ 30
hydroxyapatite	≈ 55
$CaCO_3/SiO_2/MgCO_3$ mixture	15

Bone is a complex organic–inorganic composite. The organic component, **collagen**, is a protein made up of three polypeptide chains; it has the higher-order structure[G] shown in

---

[*] The whole answer to this question is complex, involving proteins that are important in 'delivering' calcium ions to the broken area.

Figure 6. Carboxylate-containing amino acid side-chains are regularly spaced along the collagen fibres.

☐ Do carboxylate-containing side-chains act as hard or soft ligands?

■ Carboxylate coordinates through oxygen; we expect these side-chains to be hard, and therefore to coordinate to hard metal ions (Table 5).

So, as Ca^{2+} is a hard metal ion, carboxylate groups should act as good ligands for it. This is part of the chemical process that selectively deposits calcium ions in bone. However, this is only part of the answer. Remarkably — and this fact is still poorly understood — the spacing of carboxylate groups on collagen is such that the regular distance between neighbouring coordinated calcium ions, is *the same* as that found in crystals of hydroxyapatite, $Ca_{10}(PO_4)_6(OH)_2$ (see Section 2). (Remember that a crystal contains a lattice of regularly arranged atoms/molecules.) As a result, small crystals of hydroxyapatite grow exclusively on the collagen fibres, resulting in the formation of bone (Figure 7).

☐ What might you liken this process to?

■ This phenomenon is the same as the purification of a compound by recrystallisation. Once small crystals of a particular compound begin to grow, other molecules of the same compound crystallise onto the small crystals — even from a solution containing many different molecules — and so the crystals grow.

Figure 6 The higher-order structure of collagen, showing three intertwined helical polypeptides.

Figure 7 Schematic diagram of the biomineralisation process in bone. The regular spacing of the carboxylate amino acid side-chains along the collagen polypeptide strand gives rise to regularly spaced calcium ions. The spacing is similar to that observed in single crystals of hydroxyapatite, and so promotes the 'growth' of small crystals of hydroxyapatite on the collagen strand. Grey lines indicate ionic bonding.

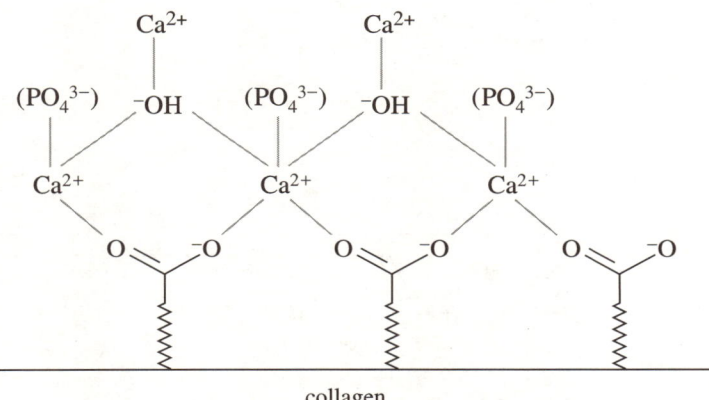

Therefore, using a combination of a hard ligand with a hard metal ion, and a rigid, regularly spaced array of ligands, collagen is able to chelate calcium ions selectively in the bone-growing process. The whole process is known as **biomineralisation**[G]. (Very similar chemical processes are involved in shell growth.) Biomineralisation is a good example of how amino acid side-chains attached to a rigid protein structure can lead to extraordinary metal coordination chemistry: we shall see this again and again throughout this Block.

The concept that ligands in a fixed arrangement chelate a metal selectively, is an important one in biochemistry. In Block 4 (Section 2.5) we saw the phenomenon in a chemical context with the synthetic crown ligands (examples are shown here in Figure 8). For example, in 12-crown-4 the crown size is so small that it chelates the small Li^+ ion very effectively. Other ions, like Na^+ and K^+, although chemically very similar, are too big to be 'encapsulated' by the crown ligand; compared to the Li^+–12-crown-4 complex, the analogous Na^+ and K^+ complexes are of low stability.

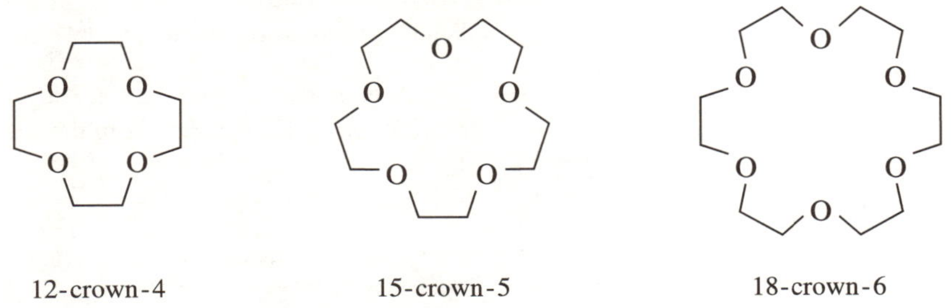

Figure 8 Structures of crown ligands; 12-crown-4 (selective for Li^+), 15-crown-5 (selective for Na^+), 18-crown-6 (selective for K^+). The name 12-crown-4, for example, shows that there are twelve atoms in the ring and four oxygen atoms.

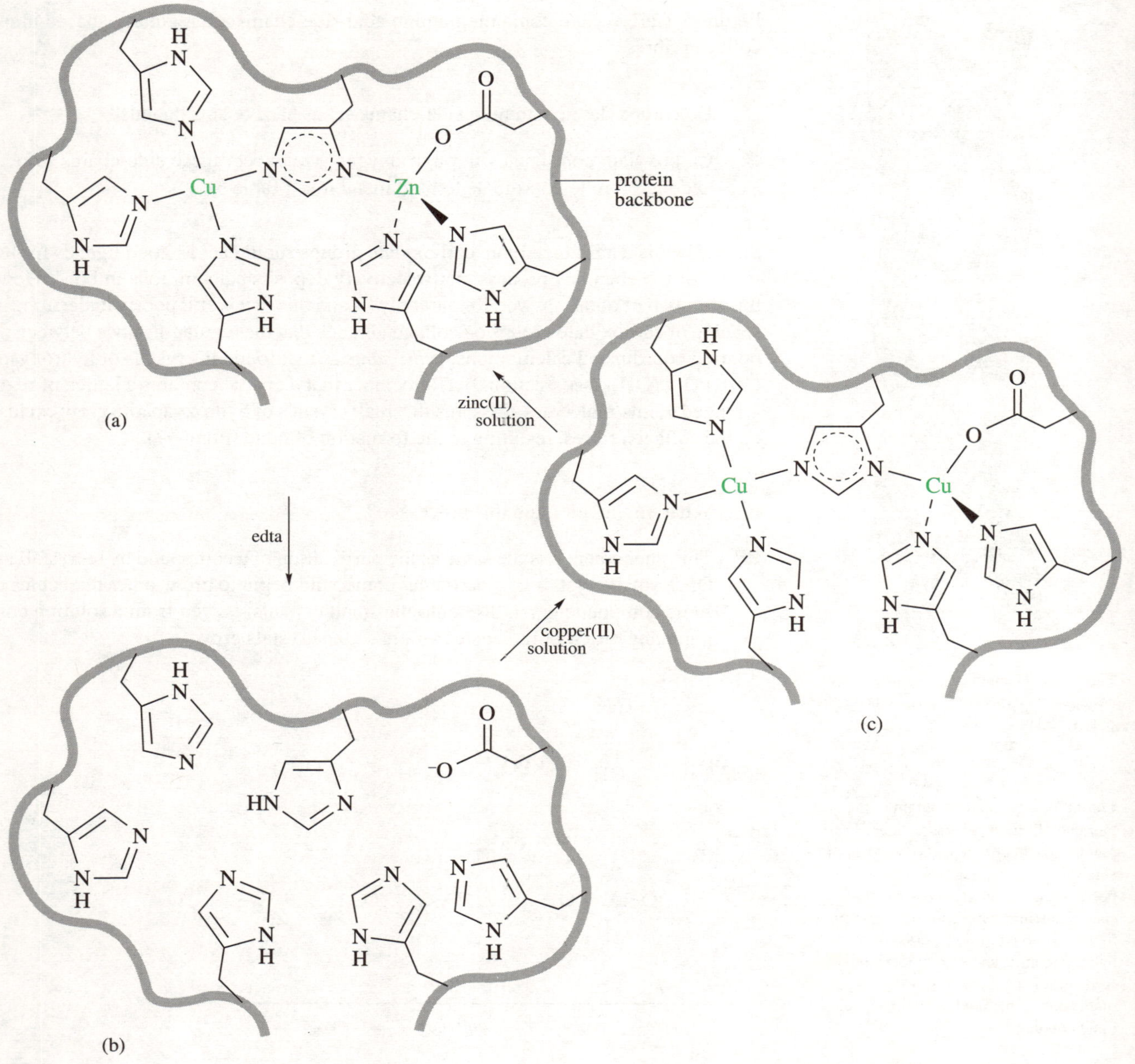

Figure 9 (a) Active site of copper–zinc superoxide dismutase showing copper and zinc sites linked with a bridging histidyl ligand; (b) treatment by edta solution to give active site of metal-free protein, showing intact amino acid structure; (c) active site after treatment of metal-free protein with aqueous copper(II) solution. (Note that the coordination geometry of the copper site is distorted square planar.)

In individual proteins, amino acid side-chains are held together in a regular, rigid arrangement by the higher-order structure of the protein chain to which they are attached. A good example of the importance of rigid structure in proteins comes from an enzyme found in human blood plasma called copper–zinc superoxide dismutase[G]. The structure of the protein's active site is shown in Figure 9a. It contains one zinc(II) ion and one copper(II) ion. The zinc(II) ion is coordinated by two neutral histidyl side-chains, one deprotonated histidyl side-chain* and one aspartyl side-chain. The copper(II) ion is coordinated by three neutral histidyl side-chains and one deprotonated histidyl side-chain.

As shown in Figure 9b, the metal ions can be removed from the protein by treating it with edta, a strong chelating agent, at pH 7; this gives an intact, metal-free protein.

☐ What happens to the deprotonated bridging histidyl under these conditions and why?

■ In the absence of metal ions, the original deprotonated histidyl side-chain is reprotonated, as expected from the pK_a value of 14 of the conjugate acid.

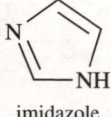
imidazole

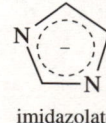
imidazolate

*The histidyl side-chain is based on the heterocyclic compound imidazole (see margin). Hence, the deprotonated side-chain ring is known as an *imidazolate*.

This demonstrates the earlier point that deprotonation can be induced by metal ions that coordinate to the deprotonated amino acid side-chain, even at a pH lower than the pK_a of the conjugate acid.

Subsequent treatment of the metal-free protein with an aqueous copper(II) solution gives a protein with copper(II) ions in *both* of the original copper(II) and zinc(II) sites (Figure 9c). Finally, treatment of the copper(II)-only protein with an aqueous zinc(II) solution replaces the copper ion in the zinc(II) site with a zinc(II) ion; this gives back the original copper(II)–zinc(II) protein.

- ☐ From the Irving–Williams series, would you predict that copper(II) would be replaced by zinc(II)?

- ■ No. In fact, what is remarkable about the replacement of copper(II) by zinc(II) is that it is contrary to the stability expected from the Irving–Williams series. As we saw in Block 4, the Irving–Williams series predicts that copper(II) complexes are more stable than the corresponding zinc(II) complexes.

Clearly, for some reason, one of the sites must have a strong preference for a zinc(II) ion.

- ☐ What do you think might be the reason for this reversal of trends, such that zinc(II) does replace copper(II) in the protein?

- ■ The answer lies in the particular spatial arrangement of the amino acid side-chains at the zinc(II) site, both the 'size' of the site and the geometry of the ligands.

The spatial arrangement of the two neutral histidyl side-chains, aspartyl side-chain and the deprotonated histidyl side-chain is such that only a particular metal ion will give a very stable complex (exactly analogous to crown ethers only coordinating a particular alkali metal ion). In this case, that metal ion is zinc(II); it has the correct size (the ionic radius of zinc(II) is 88 pm) to be coordinated effectively by the amino acids. Other metal ions, like copper(II) (ionic radius 87 pm), with slightly different ionic radii, form a less-stable complex.

Another factor in the stability of a metal ion complex is the coordination geometry of the metal ion.

- ☐ What common coordination geometries are found in zinc(II) complexes and copper(II) complexes?

- ■ Most zinc(II) complexes exhibit tetrahedral coordination geometries, whereas most copper(II) complexes exhibit either square-planar, square-pyramidal or elongated octahedral coordination geometries. From these observations, it is reasonable to assume that the most stable zinc(II) complexes have tetrahedral geometries, and that a spatial arrangement of amino acid side-chains that gives a tetrahedral coordination geometry at the metal ion, will favour the coordination of zinc(II) over copper(II) ions. Indeed, the zinc(II) site in the protein has a near-tetrahedral coordination geometry around the metal ion.

We have seen that particular proteins can chelate particular metals with appropriate amino acid side-chains in a spatially fixed arrangement, and have come some way to understanding how organisms can selectively take up metals. We shall return to metal chelation and selection in Section 7, where we shall examine in detail how iron ions (as iron(II) or iron(III)) are chelated selectively within biochemical systems.

Another feature that is influenced by the hardness of ligands is the stability of a metal ion's oxidation state. Try answering the questions in SAQ 3.

SAQ 3 Which amino acid side-chains would form stable complexes with a copper(I) ion? Which amino acids would form stable complexes with a copper(II) ion?

☐ If a copper ion were coordinated by a mixture of hard and soft amino acid side-chains (e.g. two glutamyl (hard) and two cysteinyl (soft) side-chains), which would be the more stable copper oxidation state, if either?

■ There is no simple answer to this question because there is clearly a balance here. The balance is between the relative stability of the copper oxidation states, I and II. For example, a copper ion coordinated by two methionyl (soft) and two histidyl side-chains (hard/borderline) will be expected to show little preference for either of the two oxidation states.

In biochemical systems this does occur. There are classes of copper-containing proteins that can both accept and donate an electron; in other words, the copper oxidation state can change. It turns out that the copper ion in the protein is coordinated by a mixture of hard and soft ligands, such that neither oxidation state is stabilised over the other. Moreover, in many of these proteins, the amino acid side-chains are held in a fixed spatial arrangement by the higher-order structure of the protein. This fixed arrangement gives a coordination geometry around the metal ion which is neither square planar nor tetrahedral.

☐ Which oxidation state of copper is commonly observed in square-planar coordination and why?

■ Square-planar geometry stabilises copper(II) over copper(I) because of the stabilisation conferred by Jahn–Teller distortion.

Tetrahedral geometry would stabilise copper(I) over copper(II). In fact, the coordination geometry in the protein is somewhere between square planar and tetrahedral. This is illustrated in Figure 10. Figure 10a shows square-planar geometry around a copper ion. A square-planar geometry can be converted to a near tetrahedral geometry by twisting the imaginary plane formed by the copper ion and the coordinating atoms of two adjacent ligands, so that it is at 90° to the plane formed by the copper ion and the other two adjacent ligand atoms (Figure 10c).

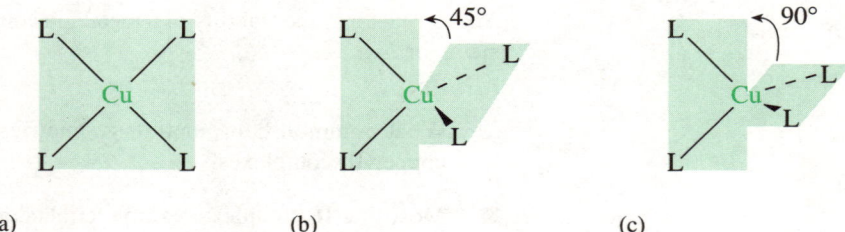

Figure 10 (a) Square-planar copper ion coordination geometry; (b) intermediate copper ion coordination geometry, showing two imaginary planes inclined at 45°; (c) tetrahedral copper ion coordination geometry, showing the imaginary planes inclined at 90°.

The copper coordination geometry found in these copper proteins can be described as a *partial* rotation of the two imaginary planes, such that the planes are inclined at an angle of about 45° to each other — in other words, half-way between tetrahedral and square-planar geometries (Figure 10b). Such a coordination geometry, which is clearly held in place by the rigid higher-order structure of the protein, does not stabilise either of the copper ion's two possible oxidation states with respect to the other. This phenomenon of a metal coordinated by a fixed spatial arrangement of amino acid side-chains is called an **entatic state**. Along with the hardness/softness mixture of the amino acid side-chains that coordinate to the metal ion, the entatic state is an example of how protein structure controls the redox potential of the metal ion within the protein.

In the case of these proteins, the entatic state and an appropriate mixture of hard/soft amino acid side-chains, give a copper ion in which the copper(II) form can accept an electron from another protein to give the copper(I) form. The copper(I) form can then be oxidised by another protein to give back the copper(II) form. It is believed that the copper protein acts as an **electron transport** protein.

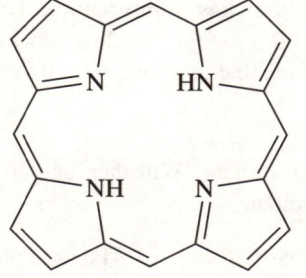

Figure 11 Porphyrin group, showing the basic tetrapyrrole structure.

3.2 Porphyrins

Metal-ion coordination by amino acids is common in biochemistry. However, other potential ligands occur in biochemical systems that are not amino acids. The most important class of these other ligands is the **tetrapyrroles**, which have the basic skeleton shown in Figure 11. (We have already seen an example of a tetrapyrrole in bioinorganic chemistry with the bacteriochlorophyll-a molecule in the previous Section.) A wide range of tetrapyrroles is known, which are distinguished from each other by the number of double bonds contained within the unit, the types of side-chain attached to the central unit, and the type of metal; see Figure 12 for some examples. The best-known tetrapyrrole is the **porphyrin group**. A form of porphyrin, as we shall see in Section 5, is an integral part of haemoglobin and the oxygen-transport protein in blood. We shall study porphyrins in detail in Section 5.

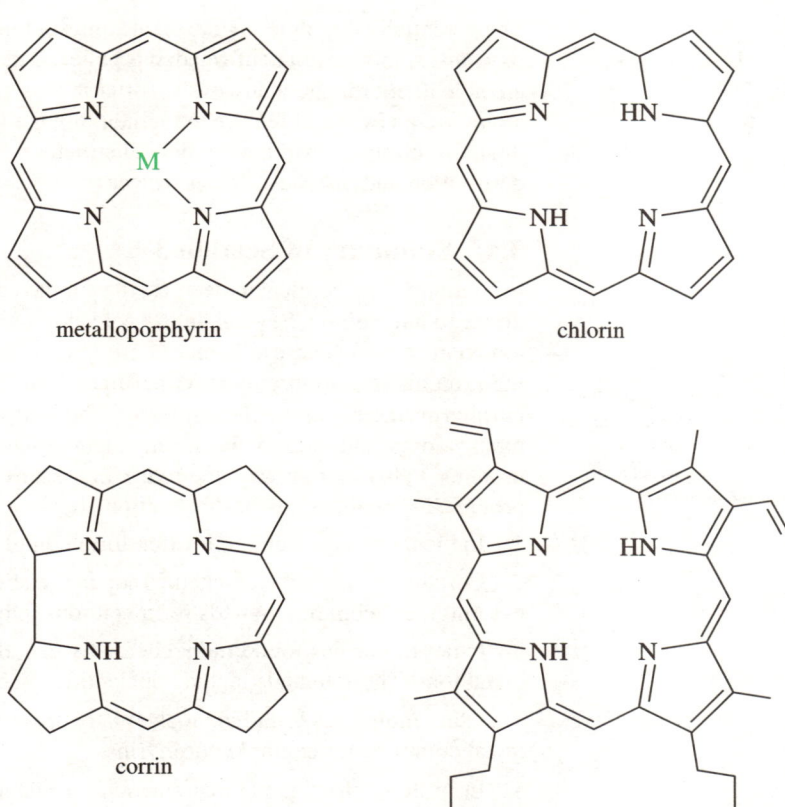

Figure 12 Various tetrapyrrole molecules.

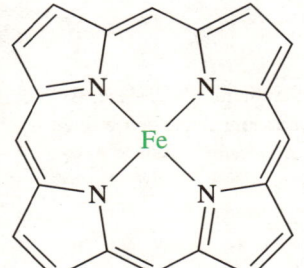

Figure 13 Iron–porphyrin group (also known as a *haem group*).

☐ What is the *hapticity* of the doubly deprotonated metalloporphyrin shown in Figure 12?

■ 4. Porphyrin acts as a tetradentate chelating-ligand for a variety of metals.

For example, Figure 13 shows a porphyrin ligand chelating an iron ion; note that two of the ligand's protons have been displaced in the complexation by iron(II) and that the porphyrin ligand has a formal charge of 2−. This is another example of a metal-ion induced deprotonation, as seen with certain amino acid side-chains.

3.3 Nucleic acids as ligands

SFC 1   Another important class of ligands in bioinorganic chemistry is the *nucleic acids*. (Remember that DNA (deoxyribonucleic acid) and RNA (ribonucleic acid) are polymers of nucleotide monomers.) Figure 14 depicts a nucleotide, which is simply a nucleobase^G connected to a ribose (sugar)/triphosphate unit. Nucleotides are the building blocks of DNA and RNA. There are several atoms within a nucleotide that in principle could coordinate to a metal ion.

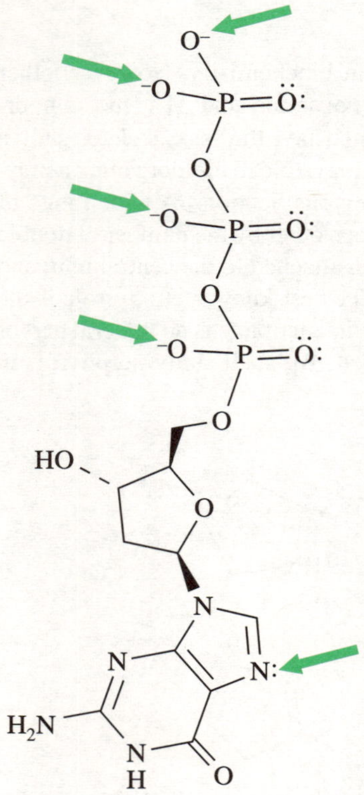

☐ Through which atom will the phosphate groups coordinate? Will they be hard or soft, and which metals do you think they will coordinate?

■ Phosphate coordinates through oxygen and so is classified as hard. The hard phosphate groups form stable complexes with hard metal ions like Ca^{2+} and Mg^{2+}. In fact, single-crystal X-ray diffraction studies suggest that Mg^{2+} coordination by phosphate — in which the phosphate groups are on the outside of the helix — stabilises (along with extensive hydrogen-bonding) the overall DNA double-helix structure.

The potentially coordinating nitrogen atoms of nucleobases are softer than the phosphate oxygens. Although not confirmed, it is believed that the nitrogen atoms of the nucleotides are able to coordinate many of the softer metals, such as Cd^{2+} and Hg^{2+}. These nitrogen atoms are an essential feature in maintaining the DNA double-helix structure: any coordination to metals disrupts the helical structure of DNA, and potentially leads to DNA destruction and diseases such as cancer.

3.4 Summary of Section 3

In summary, biochemical molecules, like amino acids, contain many atoms that can coordinate to a metal ion. Several amino acid side-chains are particularly effective at metal-ion coordination. These side-chains are often found at the active sites of metal-containing proteins (metalloproteins). Other ligands are also used in biochemical systems, the porphyrin ring being the most notable. Through a combination of hard and soft coordinating atoms, and rigid spatial arrangement of ligands, specific metals can be chelated by proteins*; also, in this way, the redox properties of the metal can be controlled by the protein. The main points of this Section are:

1 In biochemistry, metal ions often form complexes with amino acid side-chains.

2 Certain amino acid side-chains (notably, histidyl, glutamyl, aspartyl, methionyl and cysteinyl) are common ligands for metal ions in biochemistry.

3 Other molecules found in biochemistry can also form coordination complexes with metal ions. These ligands include nucleotides, saccharides, vitamins and hormones.

4 Other molecules found in biochemistry appear to have a specific role as a ligand in a metal complex, for example porphyrins.

5 In proteins the ligands that surround a metal ion are often held in a fixed spatial arrangement by the higher-order structure of the protein. This arrangement is often the key factor in determining the redox properties of the metal ion.

6 The tendency of a biochemical ligand to coordinate to a particular metal ion can be estimated using hard/soft acid–base theory.

SAQ 4 Why is Pt^{2+} a potential carcinogen?

SAQ 5 Aluminium has been implicated in possibly triggering Alzheimer's disease. By a comparison of the charge : ionic radius ratio of Al^{3+} with the bioinorganic ions listed below, identify the metal that Al^{3+} is likely to substitute in biochemical systems. Explain why.

Ion	Ionic radius/pm	Charge : radius/pm^{-1}
Al^{3+}	68	0.044
Mg^{2+}	86	0.023
Li^+	90	0.011
Na^+	116	0.009
Ca^{2+}	114	0.018
Fe^{3+}	79	0.038
Cu^{2+}	87	0.023
Zn^{2+}	88	0.023

* It is also known that the chelation of metal ions by some proteins is mediated by another protein, which may 'deliver' the metal ion or interact with the host protein.

Figure 14 Structure of guanosine triphosphate. The arrows indicate potential coordinating atoms on the molecule. Only one nitrogen atom of the nucleobase is a sufficiently strong Lewis base to act as a potential coordinating atom.

4 BIOMOLECULAR STUDY: PROBLEMS AND SOLUTIONS

Thus far in the Course, we have seen how a variety of physical techniques can be employed to elucidate the structure of small molecules. Spectroscopic techniques, such as vibrational (infrared and Raman), ultraviolet/visible and nuclear magnetic resonance spectroscopy, can be used to study features of a molecule. For example, Figure 15 shows the ^{1}H-n.m.r. spectrum of the amino acid serine. Assigning the bands in the spectrum can be achieved fairly easily from the chemical shifts and splitting patterns.

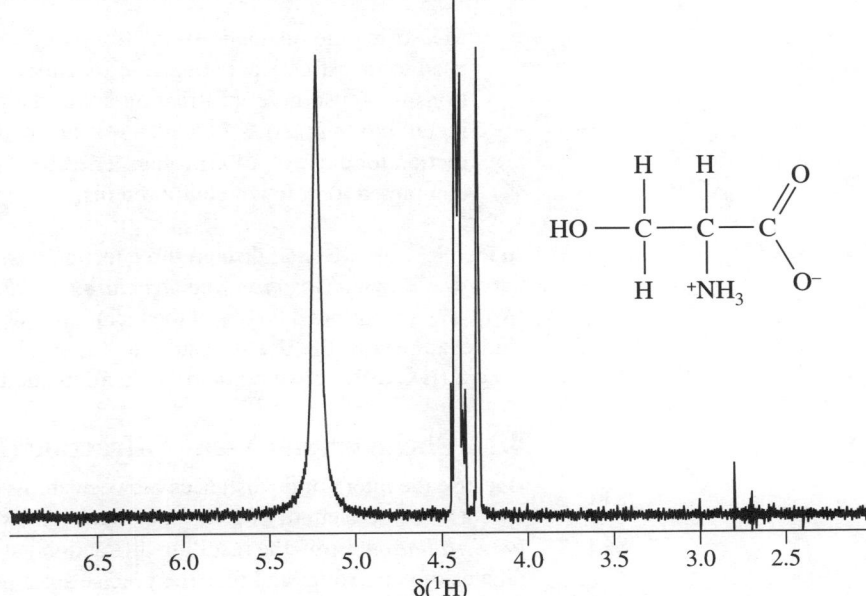

Figure 15 ^{1}H-n.m.r. spectrum of serine in 99 per cent deuterated water. There is no peak corresponding to the hydrogen atoms on the $-NH_3^+$ and OH groups, because they have exchanged with deuterons (D) in the solvent. The peak at δ 5.2 is due to H_2O in the D_2O.

When it comes to studying larger molecules, such as proteins, we encounter significant problems. Figure 16 shows the ^{1}H-n.m.r. spectrum of the protein haemoglobin, which contains *c.* 600 amino acid residues and has a relative molecular mass of *c.* 64 500. As we can see, the complex structure of the protein leads to a very complicated n.m.r. spectrum, containing many overlapping and unresolved bands: there is too much information in the spectrum, and the detail prevents us from deriving any useful information about the structure of the protein. We certainly cannot say anything about its higher-order structure or the shape and composition of its active site.

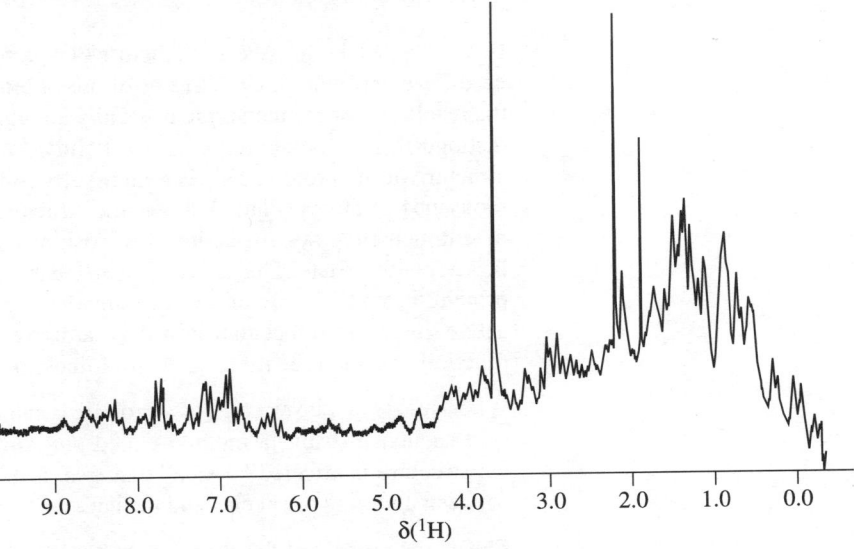

Figure 16 ^{1}H-n.m.r. spectrum of haemoglobin.

If we are to study proteins successfully, we must use physical techniques that allow us to pick out *individual* structural features of the protein, such as the nature of the active site. There are two possibilities: (i) a technique that gives us information about all of a protein, but in a form where the information can be split up, and thus interpreted and understood, or (ii) a technique that only gives us information about a *small* part of a protein — for example, the active site — thus avoiding the problem of too much information.

☐ Are there any techniques you know of, that could be used to study metalloproteins?

■ Most techniques that we have encountered so far cannot be used to give any useful information about a protein, because they suffer from the problem of yielding too much overlapping information. However, ultraviolet/visible spectroscopy can be used to study certain metalloproteins. Proteins that contain transition metal ions may show d–d or charge transfer bands in the u.v. or visible range of the spectrum. (The rest of the protein is essentially transparent to most wavelengths of visible light.) So, using u.v./visible spectroscopy, we could selectively study the metal ion and how its transitions are affected by its coordination geometry.

Measuring the magnetic susceptibility of a metalloprotein, in principle, can also be used to investigate paramagnetic transition metal ions in proteins. These investigations may also give information about the coordination geometry of the transition metal ion. In practice, however, it is difficult to obtain an accurate diamagnetic correction for the rest of the protein. Therefore, magnetic measurements have only been completed for a few metalloproteins.

TLC 1

In this Section, we shall discuss three techniques, *single-crystal X-ray diffraction (XRD)*, *extended X-ray absorption fine structure (EXAFS)* and *resonance Raman spectroscopy (RR)*. The background theory of the techniques will only be discussed briefly, but you will find examples of the useful results that they give about metalloproteins throughout the Block. (EXAFS is covered in more detail in another third level course.)

4.1 Single-crystal X-ray diffraction (XRD)

SFC 2

SLC 3

Because the interatomic distances between atoms in crystals are the same order of magnitude as the wavelength of X-rays, *crystals act as three-dimensional diffraction gratings for X-radiation*. From the resulting diffraction pattern it is possible to determine the coordinates of each atom, and thus the precise structure of molecules and their position relative to each other in the *unit cell*. Solving the structure of macromolecules is usually more difficult and time-consuming than for small molecules, simply because they have so many atoms. This can also mean that the final structure is not as well resolved as for smaller molecules. Of all of the techniques available to us, **single-crystal X-ray diffraction (XRD)** has been the one associated with the largest scientific advances in understanding protein structure and function. The starting point for XRD is to isolate and crystallise the protein under study. Both these procedures — isolation and crystallisation — require great practical skill and are very time consuming. Plate 2 (back cover) shows crystals of a protein under a microscope. The crystals are about 1 mm long and are suitable for analysis by XRD.

In essence, XRD can give a full picture of a protein's structure and, in the most favourable cases, even provides the relative positions of individual atoms. In protein crystallography, the resolution of the technique is usually about 200 pm, which is usually good enough to distinguish individual atoms: it is certainly good enough to determine the higher-order structure of the protein. This is remarkable, considering that even small proteins contain thousands of atoms. Plate 3 shows the structure of carbonic anhydrase (an enzyme), as determined by X-ray diffraction. The positions of individual atoms within the protein can be seen in this Plate. The active site can clearly be seen as a large 'cleft' in the side of the protein. From this structure we can identify the individual amino acids that make-up the active site and also speculate how these amino acids are involved in the enzyme's catalytic function. We shall see more of the usefulness of this structure determination in Section 8.

When studying the structure of a protein, it can be possible to alter selectively the amino acid sequence (using a method called *site-directed mutagenesis*) and then study how the structure is affected by small changes. It may also be possible to study the bonding between a variety of species and the active site.

The whole process of the structure determination of a protein can take years to complete! Despite the enormous effort required, the results from XRD are indispensable for understanding exactly how macromolecules work. We have now seen two examples of the power of the technique, that of carbonic anhydrase, above, and also the structure of the light-harvesting protein in Section 2 (Plate 1). Other famous biological structures that have been solved by this technique were the early successes with insulin and vitamin B12, to name but two. We shall meet more structures in later Sections, and in particular we shall see how important it is to know the three-dimensional structure of the active site precisely.

4.2 Extended X-ray absorption fine structure (EXAFS)

Most of us, at some time or another, have had an X-ray taken of part of our bodies. The medical use of X-rays relies on the fact that bone absorbs more X-radiation than normal tissue. X-ray absorption actually occurs in a wide range of materials. The amount of absorption is determined by the chemical composition and density of the material, and the wavelength of the X-rays. The **EXAFS** technique is dependent on the selective absorption of monochromatic X-rays by a particular element. Each element has a characteristic **X-ray absorption spectrum** and so for a metalloprotein it is possible to obtain an X-ray absorption spectrum for the particular metal that it contains.

☐ Is the X-ray region of the electromagnetic spectrum of higher or lower energy than the ultraviolet? Which transitions do you think X-rays excite in an atom?

■ X-rays have *higher* energy than ultraviolet radiation. Whereas ultraviolet can excite electrons in valence energy levels, X-rays cause electronic transitions from the inner atomic orbitals (core energy levels).

Each type of atom (and in the molecules we are concerned with in this Block, this is a metal atom) has its own individual set of core electronic levels, and so transitions between them occur at characteristic frequencies: X-rays with these particular frequencies are needed to excite these transitions. In recent years it has become possible to use a **synchrotron** to produce what is called a tunable source of X-rays; this means that you can obtain X-rays of any wavelength, within certain limits. The synchrotron operates by accelerating electrons around a large ring (≈ 30 m diameter). When they reach speeds approaching the speed of light, in the presence of a magnetic field, they start to emit radiation known as **synchrotron radiation**. Several countries now have synchrotron facilities and the one for the UK is situated at the Daresbury Laboratory near Manchester. The key property of X-ray absorption spectroscopy when it comes to studying biomolecules is that the wavelength of X-rays can be selected such that the X-rays are only absorbed by the particular metal ion within a protein; the rest of the protein is transparent to those X-rays. Hence, any information that can be obtained from the absorption of the X-rays will be related to the metal ion and its immediate surroundings.

A schematic diagram of a typical X-ray absorption spectrum of an atom is shown in Figure 17, showing the change in absorption versus the energy of the incident X-ray. Note that there are three distinct regions in the spectrum. The steeply rising section in the centre is due to the absorption of the X-ray and is known as the **absorption edge**, which occurs at a characteristic energy for each metal. Thus, for typical copper and zinc compounds, the absorption edge due to the 1s shell occurs at c. 9.0 and 9.7 eV, respectively. To the left of the absorption edge is the region known as the pre-edge, with a steeply falling background due to the scattering of the X-rays by matter; this need not concern us here. To the right of the edge is the EXAFS region, also characterised by a steeply falling background, but on which are superimposed some oscillations known as the fine structure. It is these oscillations which are of value for the study of metalloproteins.

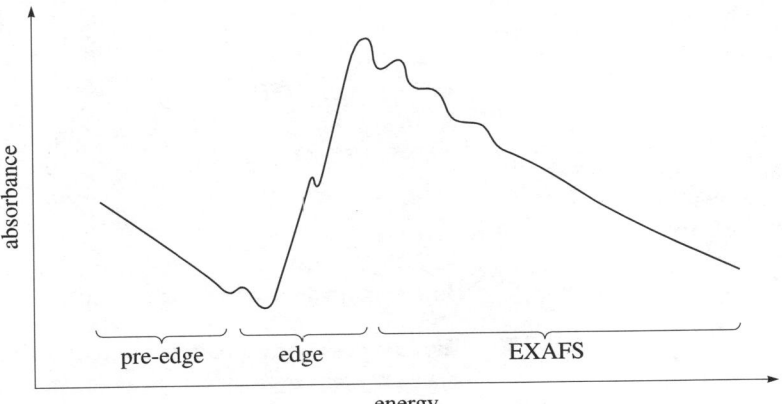

Figure 17 A typical X-ray absorption spectrum.

Figure 18 shows us how these oscillations can be understood. The X-ray region to the right of the absorption edge has sufficient energy to eject a core electron. Averaged over the whole sample, electrons are emitted in all directions. We can understand the EXAFS phenomenon in terms of a simple model. The emitted electrons can be considered as a spherical wave moving out from the metal atom (Figure 18a). The metal atom is surrounded by other atoms of the structure, and this outgoing wave will meet these and be reflected by them (Figure 18b and c). The backscattered waves will now meet the outgoing wave and interfere with it: the *interference* can be either constructive or destructive depending on the distance of the backscattering atom from the metal. The extent to which the outgoing wave is reflected by another atom and, therefore, the intensity of the reflected wave, depends on several factors, including the atomic number of the backscattering atom. The final interference pattern is thus characteristic of both the *number* and the *type* of atoms surrounding the absorbing metal atom as well as their *distance* from the metal.

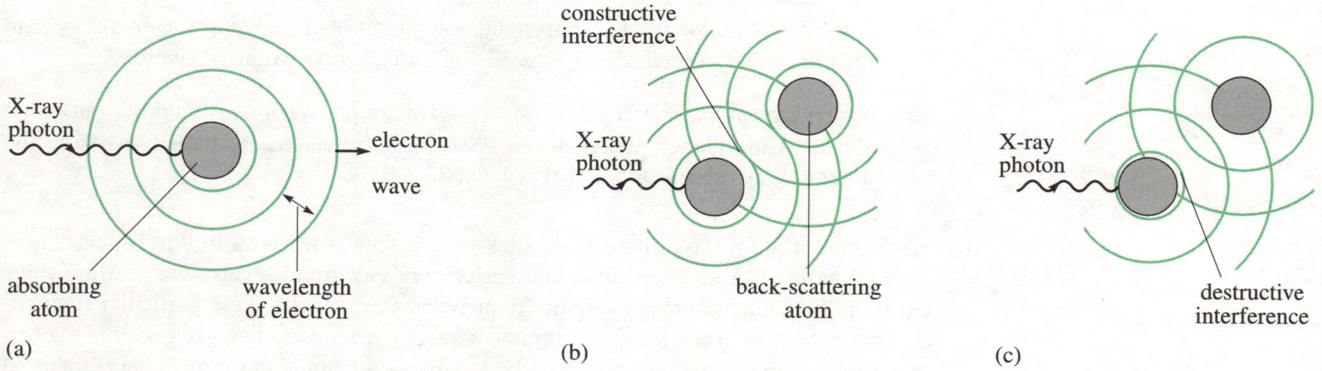

Figure 18 EXAFS can be modelled by the following processes: (a) the generation of an electron wave by ionisation; (b) constructive interference in scattering of the electron wave; (c) destructive interference.

Suitable computation of these data, involving background subtraction, Fourier transformation and refinement, eventually lead to a plot known as a **radial distribution function**. Essentially, EXAFS enables you to refine the amount of electron density as a function of distance away from the absorbing metal ion, which can extend out to about 600 pm away from the metal ion in favourable cases (remember that this distance is for all directions around the metal ion). It is important to note that *only* distance information is available: no angular information is obtained. Therefore, the technique gives us information about the immediate coordination shell of the metal ion. In particular, it gives us information about the number and types of atom in the metal ion's coordination sphere and their distance from the metal (although it is often difficult to tell apart atoms with similar atomic numbers). This information may, in turn, be extrapolated in order to establish the types of amino acid side-chains that are coordinated to the metal ion.

Let's take $[Co(CO)_4]^-$ as a simple example of the EXAFS technique. The structure of this ion is predicted to be tetrahedral, as shown in Figure 19a. An EXAFS experiment based on absorption at the cobalt edge would therefore be expected to show four carbon atoms at equal distances from cobalt, with four oxygen atoms further away, again all at the same distance from cobalt. The radial distribution plot of $[Co(CO)_4]^-$ from a Co-EXAFS experiment is shown in Figure 19b. The plot shows two peaks, one centred at 177 pm and the other at 292 pm.

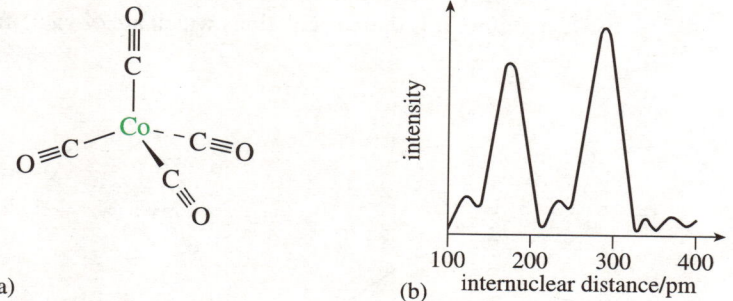

Figure 19 (a) Molecular structure of $[Co(CO)_4]^-$; (b) Co-EXAFS radial distribution curve of $[Co(CO)_4]^-$.

☐ How would you interpret these data?

■ The first peak corresponds to the four carbon atoms, which are all at the same distance (177 pm) from the cobalt atom. The second peak, at 292 pm, is due to the four carbonyl oxygen atoms.

These data thus confirm the structure shown in Figure 19a, and show that cobalt is surrounded by two shells of neighbouring atoms known as **coordination shells**.

☐ Figure 20 indicates the coordination geometry of the zinc ion in carbonic anhydrase. (This version of the protein has been doped with iodide, so that an iodide anion is coordinated to the zinc ion in place of a water ligand in the natural enzyme.) Predict the form of the radial distribution function that would be obtained from a Zn-EXAFS experiment.

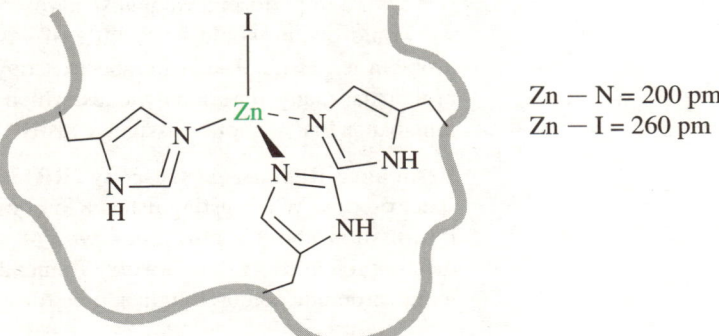

Zn — N = 200 pm
Zn — I = 260 pm

Figure 20 Part of the active site structure of carbonic anhydrase. (In the natural enzyme the iodide site is occupied by a water ligand.)

■ As all the non-iodide atoms that are directly attached to the zinc (three nitrogen atoms of three histidyl side-chains) are about 200 pm from the zinc (Figure 21a), we expect a peak in the radial distribution plot at about 200 pm (Figure 21b). The Zn—I distance is somewhat longer (260 pm), and so we would expect to observe a second peak at about 260 pm. (Although there is only one iodide ligand, the iodide is a strong backscatterer since it has many electrons. We would therefore expect this peak to be strong.)

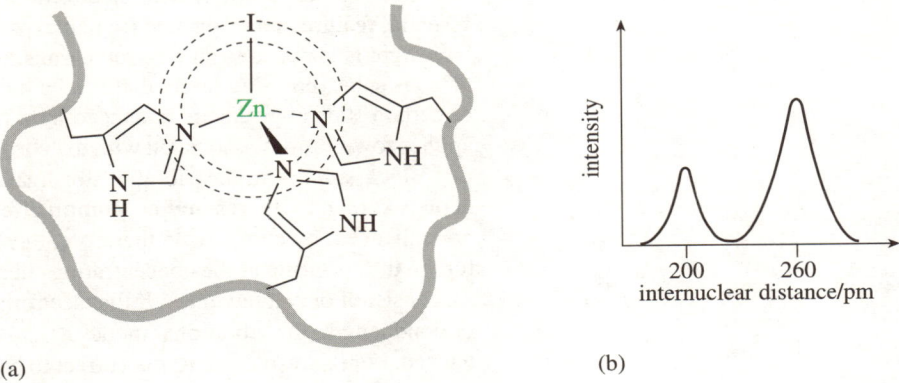

Figure 21 (a) Active site structure of carbonic anhydrase, with circles around the zinc atom of radii 200 pm and 260 pm; (b) schematic Zn-EXAFS radial distribution plot for the iodide form of carbonic anhydrase.

Therefore, from an EXAFS experiment it is possible, in favourable circumstances, to determine the number, type and distance of the atoms that are directly coordinated to the target metal atom. In the absence of a full crystal structure, this is very valuable information to be gleaned about a metalloprotein.

EXAFS has one big advantage: namely, single crystals of the protein do not need to be grown; powdered, or even solution, samples can be used. Its two disadvantages are that no angular information is available, and it is only reliable up to about 600 pm away from the metal atom. It also has the practical disadvantage that a very high-energy, tunable X-ray source is needed to perform the experiments, and only a few of these exist. Nevertheless, EXAFS is widely used in bioinorganic chemistry and we shall encounter it again later in the Block.

4.3 Resonance Raman spectroscopy

SLC 4 You have already encountered *Raman spectroscopy* in a second level course.

☐ What transitions are studied by Raman spectroscopy?

■ Raman spectroscopy gives us information about the *vibrations* of molecules; it is a complementary technique to infrared spectroscopy.

In Raman spectroscopy an intense, coloured, monochromatic light source (usually a laser) is scattered by the sample. When this scattered light is analysed, it is found that most of the scattered light is unchanged in frequency; this is Rayleigh scattering. However, a small proportion of the scattered light is changed in wavelength from the incident light, and this is known as the Raman scattering. The incident beam is changed by amounts corresponding to the energy of certain molecular vibrations (called the normal modes of vibration), giving rise to Stokes lines below the frequency of the Rayleigh line and anti-Stokes lines at corresponding frequencies above the Rayleigh line.

If we were to perform a Raman spectroscopy experiment on a protein, the information we would obtain would be of little, if any, use. The technique suffers from the same problem as ^{1}H n.m.r. and i.r. spectroscopy of proteins (see Figure 16), in so far as a protein has many vibrational modes, which give rise to spectra with unresolved, overlapping bands that cannot be assigned easily.

Resonance Raman spectroscopy (RR) is a technique very closely related to Raman spectroscopy. What distinguishes RR from Raman spectroscopy is that RR only gives information about a *particular part* of a molecule, or for our purposes, part of a metalloprotein. How does it work? Remember that Raman spectroscopy uses an intense, monochromatic coloured light source for its experiments.

☐   What other feature of a metalloprotein might be affected by coloured light?

■   Metalloproteins often have a transition metal as part of the active site, and transition metals may absorb visible light. Light of a *particular* wavelength is selectively absorbed by a transition metal ion when an electronic transition is excited.

In a normal Raman experiment, the frequency of the laser light does not coincide with, and therefore does not excite, electronic absorptions in the molecule. However, the essential feature of a resonance Raman experiment is that the wavelength of the incident laser light is matched to an electronic transition of the transition metal ion in the sample. (This is achieved using lasers that can be tuned over a particular range.) The magnitude of certain Raman peaks are enhanced by a factor of 10^2 to 10^6. The enhancement only applies to vibrations associated with the chromophore, and in the examples referred to in this Block, applies to the vibrations of ligands attached to the metal ion. The phenomenon is known as the **resonance Raman effect**. The other vibrations of the molecule are not enhanced; by comparison therefore, these vibrations are very weak, and do not feature in the spectrum at the concentrations normally used. Hence resonance Raman spectra consist of only a few lines. If the electronic transition produces a change in geometry or bond length, the vibrational modes associated with those changes are particularly enhanced. For the resonance Raman effect to be observed, the electronic transition must be symmetry allowed; the enhanced transitions are nearly always totally symmetric modes.

To take a simple example: the $[Re_2F_8]^{2-}$ ion contains a Re—Re quadruple bond, and the eight fluorine atoms are situated almost at the corners of an imaginary cube. $[Re_2F_8]^{2-}$ has an electronic absorption band in the visible/near u.v. region. When $[Re_2F_8]^{2-}$ is irradiated by a laser of the correct frequency, the resonance Raman spectrum shown in Figure 22a is obtained. The Raman effect is associated with the excitation of an electron to an antibonding level (Figure 22b), resulting in a weakening of the Re—Re bond. Note that RR is a scattering phenomenon, and *not* an absorption phenomenon, so an excited electronic state is not formed in the usual sense.

☐   What would you expect to happen to the Re—Re bond length?

■   It will lengthen as the bond order decreases.

As the electronic transition affects the length of the Re—Re bond, we might expect that vibrations associated with the Re—Re bond would be enhanced in the resonance Raman spectrum. And indeed this is the case. The resonance Raman spectrum shows a considerably enhanced intensity for the Re—Re symmetric stretch, v_1. (The progression of peaks that we see are overtones of v_1 — that is, $2v_1$, $3v_1$…, etc. — and is due to transitions between excited vibrational energy levels.)

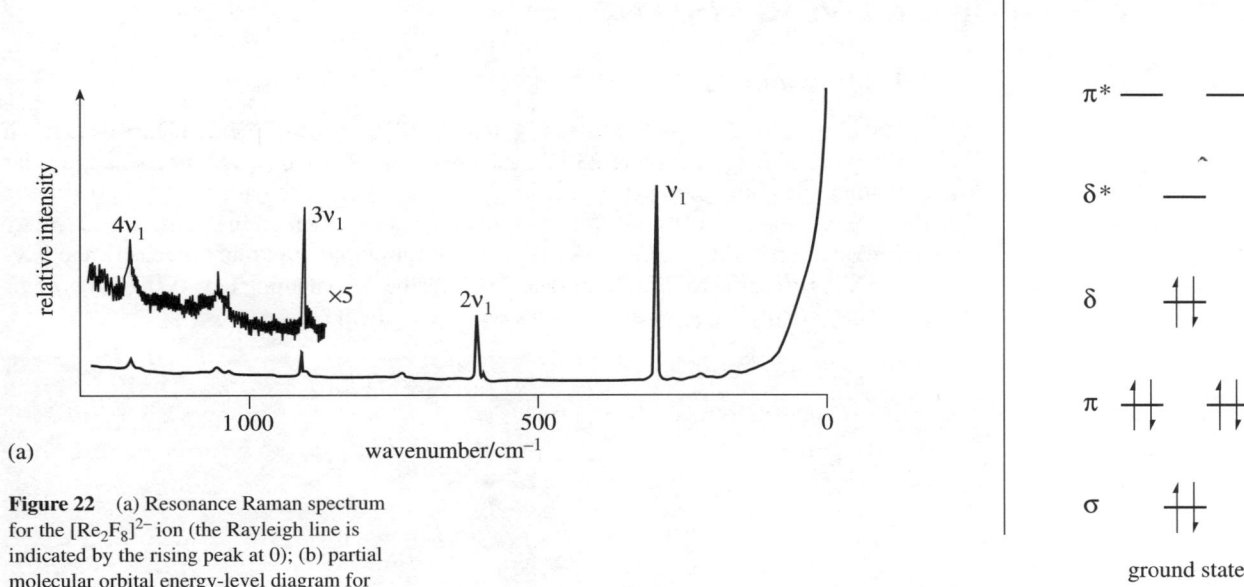

Figure 22 (a) Resonance Raman spectrum for the $[Re_2F_8]^{2-}$ ion (the Rayleigh line is indicated by the rising peak at 0); (b) partial molecular orbital energy-level diagram for the $[Re_2F_8]^{2-}$ ion in its ground state.

Unfortunately, RR experiments are difficult to perform. For instance, the sample may have to be rotated in order to reduce the effects of local heating induced by the absorbed laser light, because this can cause the sample to decompose. So, not all metalloproteins can be studied in this way. Despite this, the information available is valuable in identifying the ligands that are attached to the metal; in many cases, these also include the protein's substrate itself. We shall see examples of this in Section 5, where we shall examine proteins that bind molecular oxygen. Resonance Raman spectroscopy has been used to observe the O—O vibration of O_2 bound to the metal active site of these proteins.

4.4 Summary of Section 4

In this Section we have discussed three physical techniques that are used in bioinorganic chemistry.

1 Single-crystal X-ray diffraction is the most powerful technique, in that it gives a full three-dimensional picture of the protein, often with individual atomic positions.

2 EXAFS gives information on the number, type and interatomic distance for atoms surrounding a particular metal centre. The coordination shells can be identified up to about 600 pm in the best cases. It provides no angular information.

3 In resonance Raman spectroscopy the intensity of some of the vibrations of bonds between a metal chromophore and its ligands are enhanced by tuning the incident laser to an electronic absorption frequency of the metal ion. The spectra are simpler than a conventional Raman spectrum because they have fewer bands. The vibrational bands give information about the coordination around the metal centre.

4 Resonance Raman spectroscopy and EXAFS do not give as much, or as precise, information as XRD, but, rather, give information about the coordination of the metal centre in a protein. As the metal centre is nearly always at the active site of the protein, this information is valuable in determining how the protein functions.

5 Data from all three techniques are often sufficient to determine a detailed picture of how a protein operates. We shall see how these techniques are used later in the Block.

SAQ 6 What information is available from a resonance Raman spectroscopy experiment?

SAQ 7 Why are techniques like infrared and conventional Raman spectroscopy of limited use in studying proteins?

SAQ 8 Name one advantage and two disadvantages of using XRD to investigate protein structures.

5 OXYGEN TRANSPORT IN LIVING SYSTEMS

5.1 Introduction

On 29 May 1953, Tenzing Norgay and Edmund Hillary became the first humans to reach the summit of Everest (Figure 23). At such high altitudes the two climbers had to wear breathing apparatus (at 8 840 m, the height of the summit, the atmospheric pressure is roughly one-third that at sea-level). Without the aid of the breathing apparatus they may never have reached the summit. Air is essential for human life; more precisely, the oxygen in air is essential for life. As humans, we breathe in something like 45 000 litres of air every day simply to provide our bodies with a supply of oxygen.

Figure 23 Edmund Hillary and Tenzing Norgay climbing Everest. Without the aid of breathing apparatus the climbers may not have been able to reach the top.

Oxygen, O_2, comprises about 21 per cent (volume %) of the air we breathe. It is transported via the bloodstream from the lungs to all parts of our bodies. The oxygen diffuses from the bloodstream into the cells, where it is used in aerobic respiration, the process that provides energy. Using glucose as an energy source, the overall reaction for aerobic respiration is:

$$C_6H_{12}O_6 + 6O_2 = 6CO_2 + 6H_2O + \text{energy} \qquad \textbf{10}$$

Six moles of oxygen are consumed for every mole of glucose, and a good supply of oxygen is essential to enable our cells, and ultimately our bodies, to operate and function normally. However, not only humans require oxygen; the life of most organisms, from the smallest single-cell amoeba to the largest elephant also depends on supplies of this simple molecule.

For small, single-cell organisms, oxygen is simply and easily obtained. These organisms make use of the facts that oxygen is slightly soluble in water and that it is a small molecule, which can quickly penetrate or **diffuse**G through cell membranes.

As single cells are very small (ranging from 1 to 100 μm in diameter), the diffusion of oxygen into the centre of the cell is fast enough to support the respiration reactions; this is known as **passive diffusion**. However, the amount of oxygen which can diffuse passively through the cell drops off rapidly with the distance over which the oxygen has diffused. What this means in practice is that organisms that rely on the passive diffusion of oxygen as their source of oxygen cannot be larger than about 1 mm in diameter; for larger organisms the oxygen would not get through in large enough quantities to support respiration.

Temperature is also important. The solubility of oxygen in water *falls* with increasing temperature. At 5 °C the solubility of oxygen in water is about 2 mmol l^{-1}, which is enough oxygen in solution to maintain the respiration rate of a unicellular organism. Thus, very small organisms living at temperatures of about 5 °C are able to obtain their oxygen requirement by passive diffusion. However, at 40 °C the solubility is only around 1 mmol l^{-1}.

Two immediate problems must be overcome to satisfy the oxygen requirements of large organisms (like humans), namely:

- the rate of passive diffusion of oxygen through respiring tissue (e.g. skin) is not fast enough to penetrate much further than about 1 mm;
- the solubility of oxygen drops off with increasing temperature. The solubility of oxygen in blood plasma (the fluid component of blood, which does not contain red blood cells) at 37 °C is 0.3 mmol l^{-1} (in other words, only 15 per cent of that for pure water). So, for warm-blooded organisms, like humans, the solubility of oxygen in blood plasma is not high enough to support aerobic respiration in the cells.

To survive, large animals (that is, greater than 1 mm in size) must have a means of capturing oxygen from the air, circulating it around the animal and, if they are warm-blooded or exist in hot climates, find a way of concentrating oxygen within their circulation systems.

The first problem of circulation is largely a mechanical one; a pump and pipes are required. Of course, these are the heart and blood vessels. The second problem of increasing the concentration of oxygen within circulation systems is largely a chemical one. It is this problem and the biochemical systems that overcome it, which will be the main subjects of this Section.

As a final thought before we address the oxygen-concentration problem in detail, consider the Antarctic ice-fish! This fish has a heart and circulation system similar to all vertebrates. However, it has no means of concentrating oxygen in its bloodstream (in fact, its blood is completely colourless!) These fish live in temperatures of about −1 °C.

☐ Why does the ice-fish have no biochemical oxygen concentration system?

■ At these temperatures the solubility of oxygen in water (or colourless blood!) is about 5 mmol l^{-1}. This is high enough to support respiration in the cells of the fish, so it has no need of a chemical system to concentrate oxygen in its bloodstream.

5.2 The chemistry of oxygen

To begin with, we need to examine the chemistry and properties of oxygen in some detail. O_2 is a very reactive molecule. Nearly all the reactions of organic compounds with oxygen result in oxidation of the compound and reduction of oxygen to water. This type of reaction is nearly always highly exothermic. For example, fire is the oxidation by oxygen of organic matter (wood, oil, coal, etc.). The amount of energy given out from such a reaction is obvious.

The reason for the high reactivity of oxygen stems from its strongly oxidising nature. The reduction of oxygen to water is very energetically favourable (Table 7). O_2 can actually be reduced not only to water but also to a variety of oxygen-containing molecules and radicals. Some of these molecules are shown in Figure 24; note that the molecule formed is dependent on the pH of the solution.

The properties of different oxidation states of the O-containing molecule can be illustrated using a molecular orbital diagram.

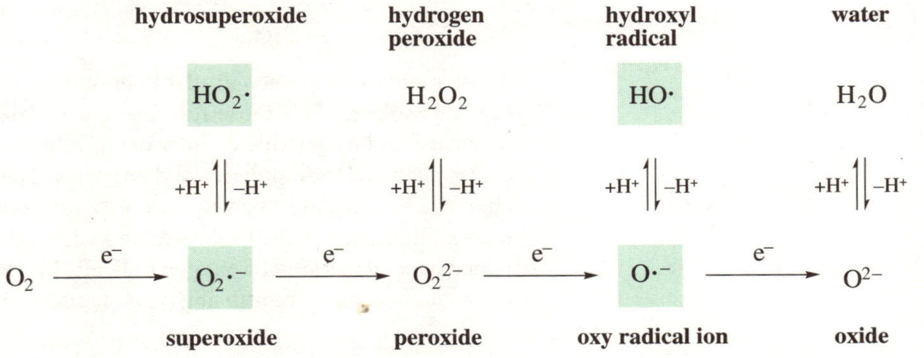

Table 7 Electrode potentials for half-reactions in aqueous solution, at pH 7, 298.15 K and 1 atm pressure

Electrode reaction	E/V
$\frac{1}{2}F_2(g) + e = F^-(aq)$	+2.89
$\frac{1}{2}Cl_2(g) + e = Cl^-(aq)$	+1.36
$\frac{1}{2}Br_2(l) + e = Br^-(aq)$	+1.08
$\frac{1}{2}I_2(s) + e = I^-(aq)$	+0.53
$\frac{1}{2}O_2(g) + 2H^+(aq) + 2e = H_2O(l)$	+0.82
$O_2(g) + H^+ + e = HO_2\cdot$	−0.34
$O_2(g) + 2H^+(aq) + 2e = H_2O_2(aq)$	+0.28

Figure 24 Species obtained on reduction of O_2 as a function of pH. Radical species are printed on a green background.

	hydrosuperoxide	hydrogen peroxide	hydroxyl radical	water
	$HO_2\cdot$	H_2O_2	$HO\cdot$	H_2O
	+H$^+$ ↕ −H$^+$	+H$^+$ ↕ −H$^+$	+H$^+$ ↕ −H$^+$	+H$^+$ ↕ −H$^+$
$O_2 \xrightarrow{e^-}$	$O_2\cdot^-$ $\xrightarrow{e^-}$	O_2^{2-} $\xrightarrow{e^-}$	$O\cdot^-$ $\xrightarrow{e^-}$	O^{2-}
	superoxide	peroxide	oxy radical ion	oxide

☐ (*Revision*) Sketch the O_2 molecular orbital diagram by combination of the atomic orbitals on two oxygen atoms. Use the molecular orbital diagram to explain why:
(a) O_2 is paramagnetic in its ground state;
(b) O_2 has a double bond and O_2^{2-} has a single bond;
(c) the vibrational frequencies decrease in the order $O_2 > O_2^- > O_2^{2-}$.

SLC 5 ■ Figure 25 shows the *molecular orbital diagram for O_2* together with pictures of the atomic and molecular orbitals. Figure 26 shows the O_2 molecular energy levels for O_2, O_2^- and O_2^{2-}, each with a different number of valence electrons. As we move from O_2 to O_2^- (superoxide^G) to O_2^{2-} (peroxide), the π antibonding orbitals gain electrons. This gain results in a gradual weakening of the O_2 bond (decrease in bond order) and a consequent reduction in the vibrational frequency of the molecule. Another result is that O_2^{2-} has a single bond compared to the double bond in O_2. (Remember that as bond strength is reduced the vibrational frequency usually decreases also.) It turns out that the vibrational frequency is a useful indicator of the formal charge on the O—O species. We shall return to this property when we come to study O—O molecules in living systems.

A closer inspection of Figure 26 shows that the superoxide anion, O_2^-, contains a single unpaired electron.

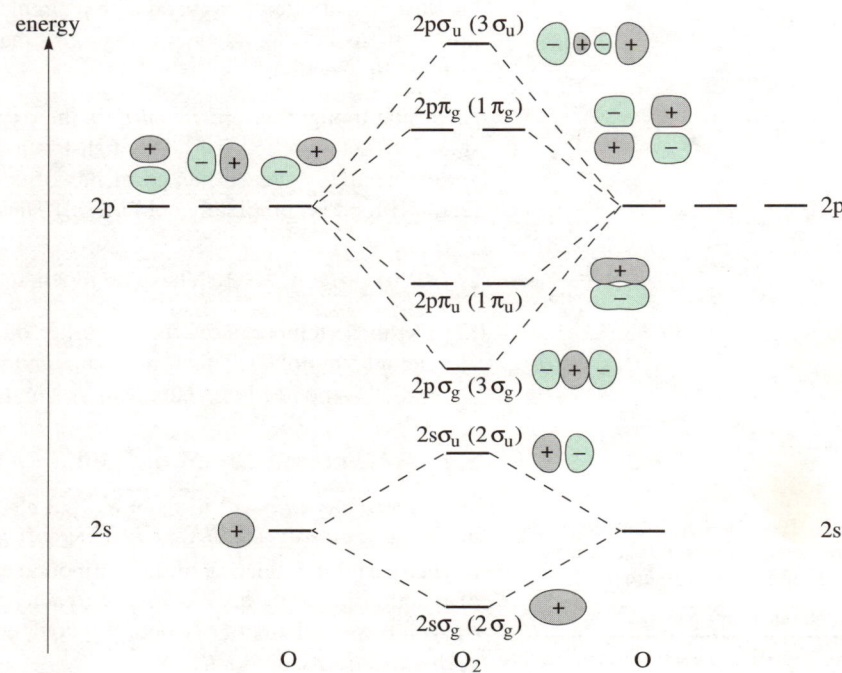

Figure 25 Molecular orbital energy-level diagram for O_2 generated by combining atomic orbitals of two oxygen atoms (for clarity, the two 1s orbitals on the oxygen atoms have been omitted). Diagrams of each type of atomic and molecular orbital are shown next to the appropriate energy level.

☐ What properties does this confer on O_2^-?

■ It is a radical, and is likely to be very reactive indeed. It is also paramagnetic.

Figure 24 also shows that, dependent on the pH, reduction of O_2 to water goes via a number of radical molecules; these are superoxide ($O_2^-\cdot$), hydrosuperoxide ($HO_2^-\cdot$), oxy radical anion ($O^-\cdot$) and the hydroxyl radical ($\cdot OH$). Furthermore, hydrogen peroxide easily cleaves to form two hydroxyl radicals in the reaction

$$H_2O_2 \longrightarrow 2\cdot OH \qquad \qquad 11$$

These radicals have uncontrollably high reactivity and will react with almost any other organic molecule*. If these reactive species are formed in living systems, they cause extensive and irreparable cellular damage by reacting with important biochemical molecules such as DNA, proteins and enzymes. Therefore, the 'handling' of oxygen in biochemical systems must be very carefully controlled so as to avoid the generation of detrimental radicals. We shall see later on in this Section that biochemical systems carefully manage the generation of oxygen radicals. In Section 6 we shall see how some enzymes are actually capable of utilising oxygen-containing radicals in biochemical reactions.

---

* Formation of radicals in living tissue is one of the consequences of irradiation with, for example, α-particles, β-particles and X-rays.

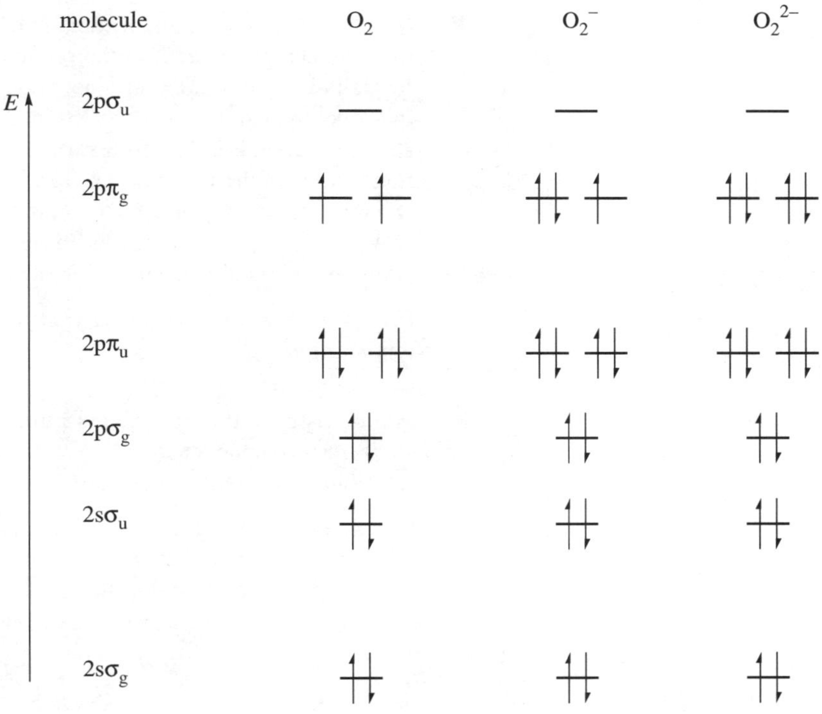

Figure 26 Molecular orbital energy-level diagram for O_2, O_2^- and O_2^{2-}, showing orbital occupancies, and comparative data for the O—O bond in each of these species. (As with Figure 25, the 1s atomic orbitals on the oxygen atoms have been omitted.)

molecule	O_2	O_2^-	O_2^{2-}
bond order	2	1.5	1
bond length/pm	121	≈ 130	≈ 140
stretching frequency/cm^{-1}	1 560	1 100 – 1 200	730 – 920

In view of the oxidising ability of O_2, it is not surprising that it reacts readily with transition metals. Transition metals, as we have seen in Blocks 1 and 3, participate in redox reactions. They exhibit a range of stable oxidation states. This property is a characteristic feature of transition metals; main-Group elements generally exhibit a narrower range of stable oxidation states. In fact, biological systems make use of this property of transition metals, and it is important for us to study briefly the types of metal–oxygen complexes that have been prepared.

Reaction of oxygen with transition metals is very common.

☐ Can you think of a common example of this?

■ We only have to remember that cars, made from steel (an alloy, mainly comprising iron in its zero oxidation state), form rust (complex hydrated iron(III) oxides) in moist air!

For the purposes of this Section we shall focus our attention on metal–oxygen complexes that exhibit *reversible* binding of O_2. In other words, we are concerned here with those complexes that will bind O_2 under high pressures of O_2, and, in their oxygenated form, release O_2 at low pressures. The first complex that was found to exhibit reversible binding of O_2 was Vaska's complex, **3**. O_2 reacts readily with Vaska's complex to give compound **4**, which contains a peroxo ligand, coordinated to the iridium in a **side-on** position:

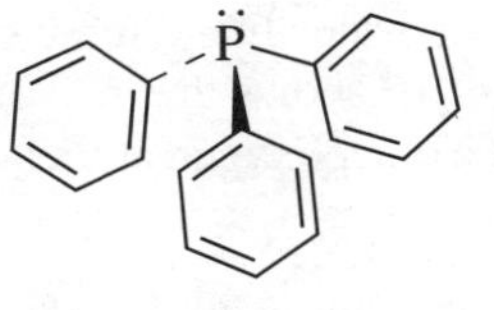

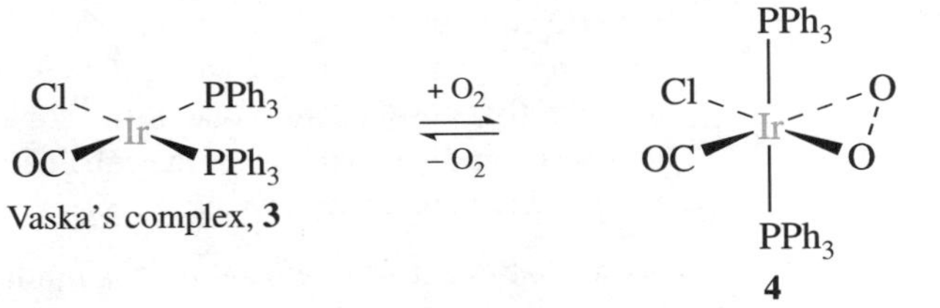

(12)

☐ (a) What are the formal oxidation states of the iridium ion in this reaction?
(b) What processes are involved in the forward and reverse reactions?

■ (a) In Vaska's complex, iridium is in a formal +1 oxidation state, because Cl⁻ is the only charged ligand. In the product of its reaction with O_2 the iridium is now formally in its +3 oxidation state, giving a peroxo ligand coordinated to the iridium in a side-on position. (It is important to note here that the assignment of oxidation state to iridium is only a formal one. It is impossible to tell exactly what its oxidation state is in the O_2 complex. The reasons for this will become clearer when we examine the Fe–O_2 molecular orbital energy-level diagram later in this Section. However, you should note that for the following examples of transition metal–O_2 complexes, it is difficult to assign the oxidation state of the metal unambiguously.)

(b) The forward reaction is *oxidative addition*. The reverse reaction is *reductive elimination*.

Removal of O_2 from the oxygenated iridium(III) complex is achieved by reducing the amount of free O_2 in the reaction vessel. In accordance with Le Chatelier's principle, the original iridium(I) complex and free O_2 are regenerated.

Another example comes from cobalt chemistry (note that cobalt and iridium are in the same Group in the Periodic Table). Reaction of $[Co(NH_3)_6]^{2+}$ with O_2 results in the reversible substitution of one of the ammine ligands by O_2. In the product, **6**, the O_2 is bound through one oxygen atom in the mode known as **end-on coordination**:

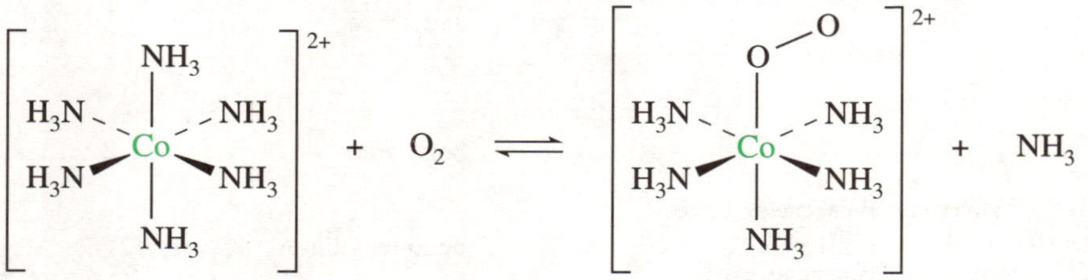

□ What is the apparent change in oxidation state of cobalt in this reaction?

■ The oxidation state change appears to be +1. $[Co(NH_3)_6]^{2+}$ contains cobalt(II). It is possible formally to envisage the Co–O_2 complex as cobalt(III) bonded to an O_2^- superoxide anion; that is, an electron has been transferred from the cobalt(II) to O_2.

A slight variation on the cobalt reaction produces an interesting result. Increasing the amount of complex with respect to oxygen leads to another compound (reaction 14). In this case the O_2 forms a bridging ligand between two cobalt pentammine moieties to give the diamagnetic molecule **7**:

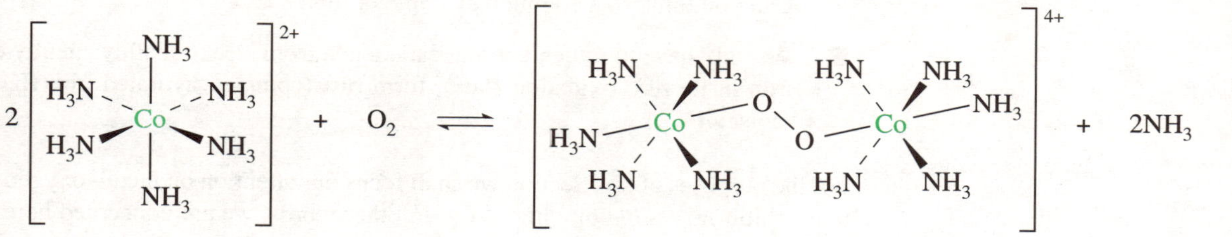

The type of O_2-binding seen in the cobalt complex **7** is known as **end-on bridging**, or μ-η^1,η^1. This nomenclature for the way a ligand bonds to a metal is widely used. There are two components in this example: μ (Greek letter mu) shows that the O_2 molecule is *bridging* the metal centres, and η (Greek letter eta), indicates the number of atoms that are directly coordinated to one of the cobalt atoms. η^1,η^1 shows that one of the oxygen atoms is coordinated to one of the cobalt atoms and the other oxygen atom is coordinated to the other cobalt atom.

As a final example of reversible O_2-binding to a transition metal, we shall consider a copper complex. Equation 15 shows the reaction of *tris*[3,5-di-isopropyl(pyrazolyl)]borate-ethanenitrilecopper(I), **8**, with O_2. (The *tris*[3,5-di-isopropyl(pyrazolyl)]borate ligand is used extensively in coordination chemistry.)

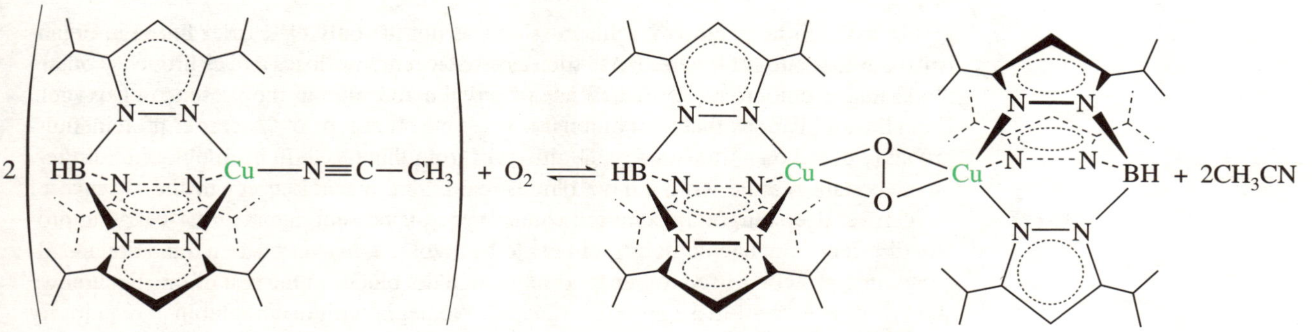

$$2\ [\text{structure } \mathbf{8}] + O_2 \rightleftharpoons [\text{structure } \mathbf{9}] + 2CH_3CN \qquad 15$$

8   **9**

The resulting O_2 complex, **9**, containing two copper atoms bridged by O_2, exhibits another mode of O_2-binding in which the oxygen coordinates in a bidentate manner to both metals. This coordination mode is known as **side-on bridging**.

☐ What will be the notation for this type of bridging?

■ $\mu\text{-}\eta^2,\eta^2$. The μ indicates that the bonding mode is bridging. η^2 indicates that a copper is bonded to both atoms of the bridge; this is true for both copper atoms.

From the examples above it can be seen that oxygen will react in a reversible manner with transition-metal complexes to give a variety of oxygen coordination modes (summarised in Figure 27).

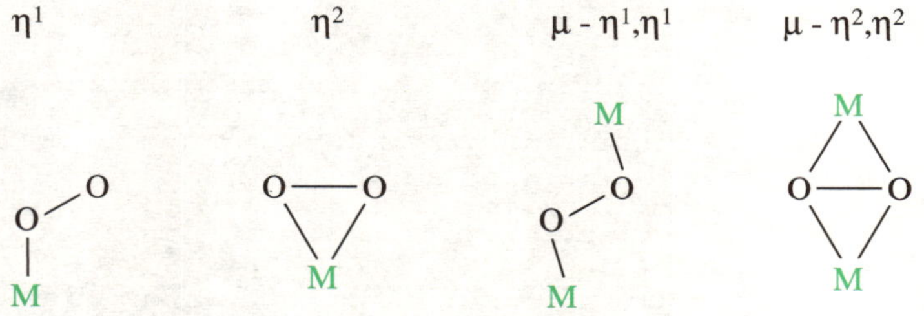

η^1 η^2 $\mu\text{-}\eta^1,\eta^1$ $\mu\text{-}\eta^2,\eta^2$

10   **11**   **12**   **13**

Figure 27 O_2 coordination modes with transition-metal complexes.

☐ Which properties of transition metals make them ideal for reversibly binding oxygen?

■ (i) A range of stable oxidation states;
(ii) a range of different coordination numbers.

We shall see that these properties are utilised in biochemical systems for oxygen transport and storage.

5.3 O_2 transport

5.3.1 Introduction

From the discussion in Section 5.1, it is clear that larger organisms must have a system for concentrating and circulating O_2 within their bodies; otherwise the passive diffusion of O_2 into the interior of the organism would be far too slow to support aerobic respiration reactions. From a chemical point of view, we also know that such organisms are likely to make use of transition metals in O_2 transport systems. The chemical properties of transition metals make them ideal centres for binding oxygen reversibly. We shall also see that another property of transition metals — the ability to form highly coloured complexes — is useful in characterising any transition metal-containing protein we study.

The brilliant red colour of blood comes directly from a chemical group called haem[G], which contains the transition metal iron. More specifically, the haem is found in the blood's O_2-carrying protein, **haemoglobin (Hb)**. Haemoglobin is present in the bloodstreams of many organisms. We shall discuss the structure and properties of haemoglobin in detail later. However, haemoglobin is not the only O_2-carrier found in organisms. Some groups of invertebrates such as crustaceans, molluscs and arthropods contain blood that is colourless in the absence of oxygen and blue in the presence of oxygen. From this, we can say that their blood utilises some other type of O_2-carrier protein, most probably based on a transition metal different from that found in haemoglobin. Furthermore, certain marine worms have bloodstreams that are colourless in the absence of oxygen yet deep burgundy coloured when oxygen is present; again an O_2-carrying protein different from haemoglobin is likely to be involved. In every case a transition metal-containing protein is used to transport oxygen in the blood. In the rest of this Section we shall discuss each of these proteins in detail. We begin with haemoglobin and a closely related protein, *myoglobin*.

5.3.2 Haemoglobin and myoglobin

Haemoglobin (Hb) is a medium-sized protein with a relative molecular mass of about 64 500. It is found in the red blood cells (erythrocytes) of many organisms, and its function is to concentrate oxygen in the bloodstream and transport it around the organism. Figure 28 shows red blood cells as seen under a scanning electron microscope. The cells are packed with Hb, which usually constitutes about 33 per cent by mass of the cell. Whole blood contains about 150 g of Hb per litre. The solubility of oxygen in blood at 37 °C is about 10 mmol l^{-1} (compare this with the solubility of oxygen at 37 °C in water, which is about 1 mmol l^{-1}, and in blood plasma, which is about 0.3 mmol l^{-1}).

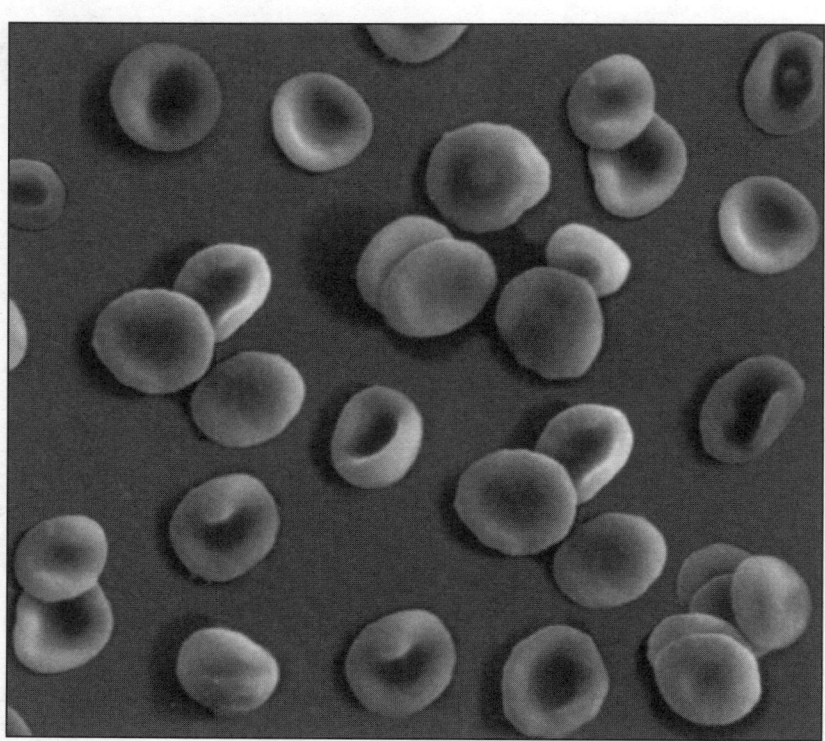

Figure 28 Scanning electron micrograph of red blood cells (erythrocytes). The cells have an unusual biconcave structure, which gives them a relatively high surface area : volume ratio. This facilitates rapid diffusion of oxygen in and out of the cell. All of the body's haemoglobin is contained in the red blood cells.

10 μm

Myoglobin (Mb), on the other hand, is a smaller protein with a relative molecular mass of about 17 800. It is found exclusively in muscle tissue, where it acts as an oxygen storage site and also facilitates the transport of oxygen through muscle. As muscles require a lot of energy to operate, they also require very efficient access to oxygen in order to maximise respiration rates.

Both Hb and Mb have been the subject of intensive study. But what we know about both proteins comes almost entirely from a knowledge of their three-dimensional structures. The structures of myoglobin and haemoglobin were determined by J. C. Kendrew and M. F. Perutz, respectively (see Box 1), using what was then the relatively new technique of single-crystal X-ray diffraction. Hb can be regarded as a Mb tetramer (that is, approximately four Mb units linked together). Therefore, we shall study the structure of Mb in some detail, before considering Hb.

Box 1 M. F. Perutz and J. C. Kendrew, the protein pioneers

1959 saw the publication of the high-resolution structure of Mb and the low-resolution structure of Hb. The story of the analysis of Hb structure is one of determination and genius, and is worth relating briefly here. M. F. Perutz (Figure 29) began his study at Cambridge University in 1936 as a graduate student. At that time, structure determination by single-crystal X-ray diffraction was in its infancy; even molecules that contained only a few atoms took many months of laborious hand calculations. Hb contains around 4 500 non-hydrogen atoms, and the solution of its structure must have seemed to be an impossible dream to Perutz and his coworkers. Perutz had to work for another 16 years before he found a means of solving the structure of Hb. This involved **isomorphous replacement**, a method of substituting in a heavy metal atom, which helps in the interpretation of the diffraction pattern: this was the turning point in protein crystallography. It then took another 6 years to obtain a low-resolution structure of the protein. Finally, in 1969, 32 years after beginning the project, Perutz determined the high-resolution structure of Hb. Perutz's work and his technique of isomorphous replacement represented perhaps the single greatest step forward in protein science this century, and revolutionised thinking about biochemistry.

Figure 29 Max Ferdinand Perutz was born in Vienna in 1914 as the son of a textile manufacturer. He began work in X-ray crystallography under J. D. Bernal in 1936 at the Cavendish Laboratory in Cambridge. His studies of the structure of haemoglobin were interrupted by six months internment as an enemy alien, and subsequently by the building of an aircraft carrier made of ice! He had previously done research into the transformation of snow into ice in glaciers. In 1947 he was appointed head of the Medical Research Council Unit for Molecular Biology, and became Chairman of the Medical Research Council Laboratory of Molecular Biology in 1962.

Figure 30 John Cowdery Kendrew was born in Oxford in 1917. After initial work in reaction kinetics, he became interested in protein structure thanks to the influence of Bernal and Sir Lawrence Bragg. His success in solving the myoglobin structure owed much to his application of computers to X-ray crystallography at a time when the structures of far simpler molecules were still unknown. He referred to his and Perutz's research as a 'mad pursuit'. In 1959, he established the *Journal of Molecular Biology*, remaining its senior editor until 1987. He was also Director-General of the European Molecular Biology Laboratory at Heidelberg from 1975 to 1982. Kendrew died in 1997.

J. C. Kendrew (Figure 30) had worked as a graduate student with Perutz, and studied the smaller protein, Mb, the structure of which could be determined more easily than that of Hb. Kendrew obtained a low-resolution (6 Å*) structure of Mb in 1957, and two years later obtained a higher resolution (2 Å) structure, from which the position of individual amino acid residues could be determined.

For their work, Perutz and Kendrew were awarded the Nobel Prize for Chemistry in 1962.

* 1 angstrom (Å) ≡ 100 pm.

5.3.3 Myoglobin structure

The higher-order structure of Mb is shown in Figure 31. The protein consists mostly of α-helices^G of polypeptide. Sandwiched between two long α-helices lies a porphyrin ring. (We have already met the porphyrin group as a ligand in Section 3.) The porphyrin in Mb has various side-chains attached to the outside of the ring. This particular derivative is known as protoporphyrin IX^G (Figure 12, p. 21). An iron(II) atom lies at the centre of the ring, coordinated by the four nitrogen atoms of the pyrrole groups; the whole unit, porphyrin and iron, is known as a **haem group**. Note that the whole haem group (apart from the side-chains) is planar (Figure 32). Mb, like Hb, has a strong red colour, which arises from π–π^* transitions within the extensively delocalised haem group. On oxygenation of deoxy-Mb the colour of the protein changes (Figure 33). This colour change identifies the haem group as the site of O_2-binding within the protein.

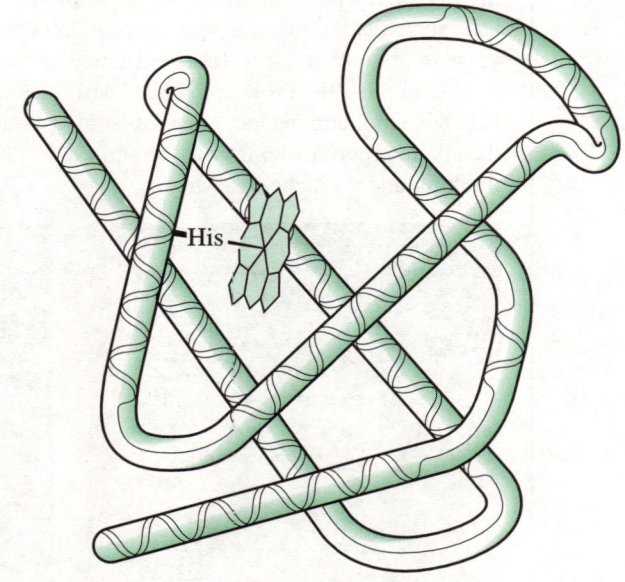

Figure 31 Schematic sketch of the higher-order structure of myoglogin, showing the large α-helical content of the protein. The haem group (see Figure 32) is sandwiched between two long α-helices within a hydrophobic pocket.

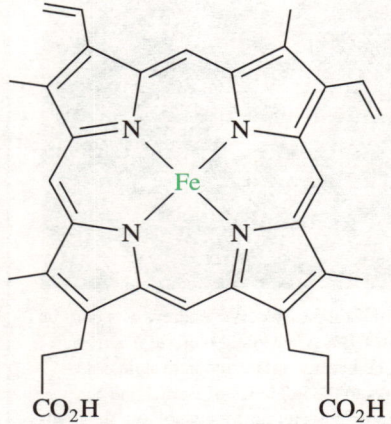

Figure 32 Haem group of protoporphyrin IX and iron. The double bonds shown are for one resonance structure; in reality the porphyrin is extensively delocalised.

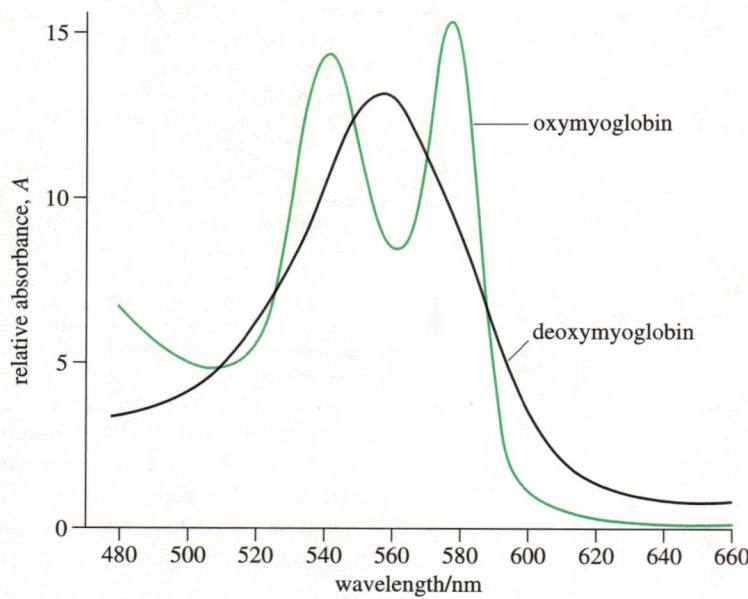

Figure 33 Visible spectra of oxymyoglobin and deoxymyoglobin. The bands are due to π–π^* transitions within the haem group. The fact that the spectrum changes on oxygenation shows that the haem group is the O_2 binding site within the protein.

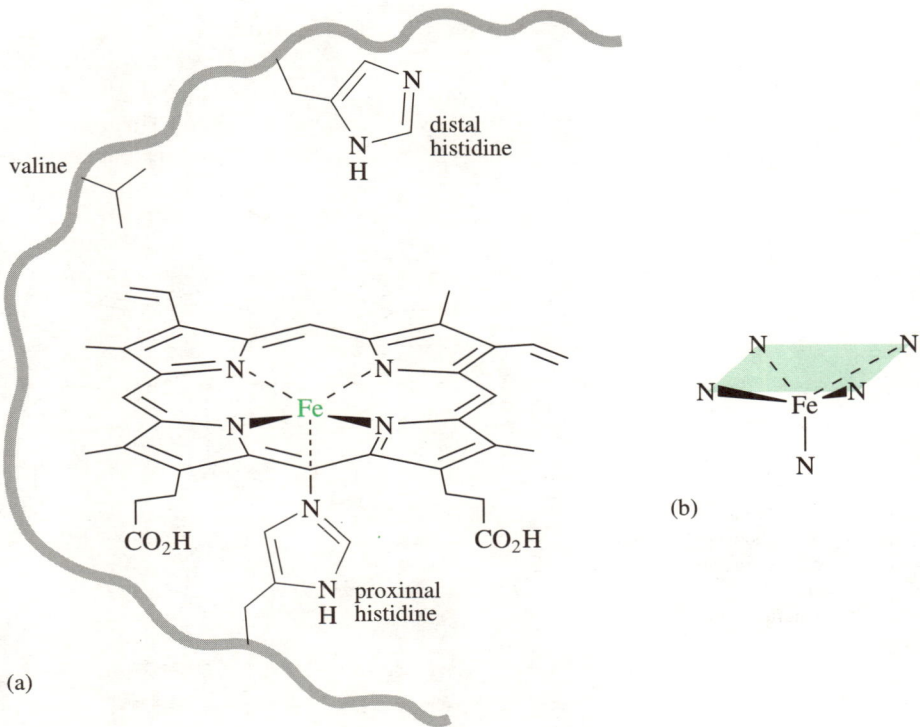

Figure 34 (a) Structure of the active site in deoxymyoglobin; (b) schematic representation of the square-pyramidal iron coordination.

A closer examination of the structure (Figure 34a) shows that the haem group is attached to the protein backbone via a histidyl side-chain. (In Section 3 we saw that histidine was one of the amino acids that could form a coordinative bond to a metal centre.) This coordinating histidyl side-chain is known as the **proximal histidine**.

☐ What is the approximate coordination geometry around the iron?

■ The histidyl side-chain and the porphyrin group give a square-pyramidal coordination geometry around the iron atom.

Figure 34b shows this coordination geometry, in which the iron atom is 'out of the plane' of the porphyrin ring, by about 60 pm. As iron(II) is usually six coordinate, the iron atom is *coordinatively unsaturated* in the deoxy form of the protein. The active site^G also contains another histidyl side-chain, which is not coordinated to the iron atom. This histidyl side-chain is on the opposite side of the haem group to the proximal histidine, and is known as the **distal histidine**; it turns out to have an important role, which we shall discuss later. The rest of the active site is taken up with hydrophobic amino acid side-chains such as valine, which prevent water from entering the active site and also prevent possible dimerisation reactions between two Mb molecules in the presence of oxygen. (See the reaction of $[Co(NH_3)_6]^{2+}$ with oxygen described in Section 5.2 as an example of such a dimerisation reaction.)

SAQ 9 How might two free haem groups be linked by an oxygen molecule? (Compare the reaction of $[Co(NH_3)_6]^{2+}$ with oxygen.)

From single-crystal X-ray diffraction studies on oxy-Mb, it was discovered that oxygen binds directly to the iron atom in an η^1-end-on fashion (Figure 35a). The further oxygen atom of the O_2 ligand participates in a hydrogen bond with the protonated nitrogen atom of the distal histidine. This extra interaction essentially stabilises the oxygen at the active site, causing it to bind preferentially over other small ligands. Binding of oxygen also results in a significant change in the iron coordination geometry. Remember that in the deoxy-form, the iron lies 60 pm out of the basal plane towards the nitrogen atom of the proximal histidine. In oxy-Mb the iron is pulled into the plane of the porphyrin ring.

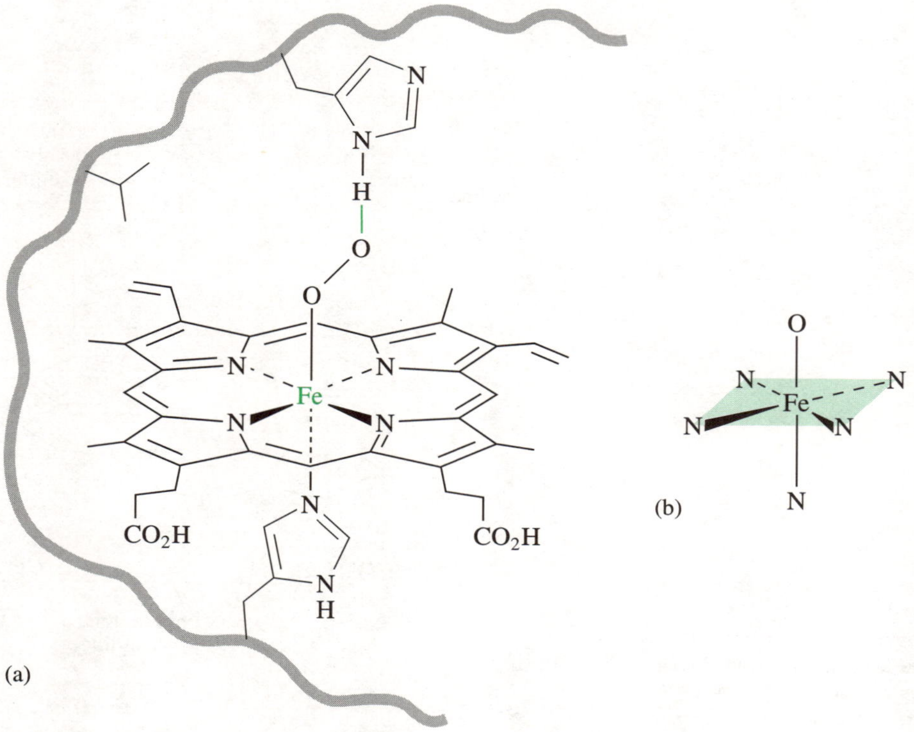

Figure 35 (a) Structure of the active site of oxymyoglobin, showing the hydrogen bond from the distal histidine to the bound oxygen molecule; (b) schematic representation of the distorted octahedral iron geometry.

☐ How would you describe the coordination around iron now?

■ In effect the coordination geometry can now be described as distorted octahedral (Figure 35b).

O_2-binding not only causes a change in structure and visible absorption spectrum of Mb (see Figure 33), but it also causes a change in the number of unpaired electrons in the haem group. Although we said in Section 4 that in general no useful information could be obtained from magnetic studies of biological macromolecules, that is not the case for Mb. An accurate diamagnetic correction is known, which allowed reliable magnetic measurements to be performed. They showed that in the deoxy state, Mb is paramagnetic, with four unpaired electrons in the haem group, whereas in the oxy state Mb is *diamagnetic* with no unpaired electrons. To understand this we need to apply a little molecular orbital theory.

☐ First, however, use crystal-field theory to sketch the d-orbital energy levels for a 3d transition metal in an octahedral crystal field. Show how the relative energies of the d orbitals change as the octahedral field is changed to a square-pyramidal field. Assume that the fourfold axis is the z axis. Feed in the appropriate number of electrons for iron(II) in both diagrams, assuming a weak-field situation.

■ The sketch in Figure 36 shows that on distortion of an octahedral field to a square-pyramidal one, the d_{z^2} orbital energy falls relative to that of $d_{x^2-y^2}$. Also the degeneracy of the d_{xy} orbital with the d_{xz} and d_{yz} orbitals is removed in that d_{xy} moves to a higher energy relative to d_{xz} and d_{yz}.

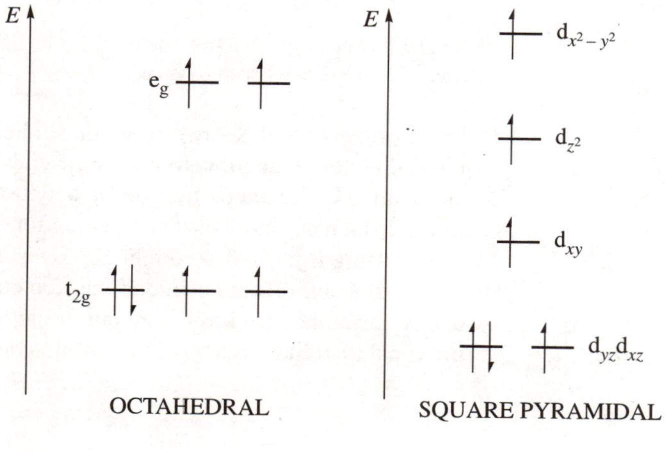

Figure 36 d-Orbital energy-level diagrams for high-spin octahedral and square-pyramidal iron(II).

The four unpaired electrons for deoxy-Mb can be explained fairly easily by reference to the d-orbital energy-level diagram for iron(II) in square-pyramidal coordination geometry (Figure 36b). Iron(II) has six d electrons; if we assume that it is in the *high-spin configuration*, then there will be four unpaired electrons.

To explain the diamagnetism of oxy-Mb we need to examine the orbitals that interact with each other when O_2 binds to the iron atom. O_2 binds η^1 end-on to iron with an Fe—O—O angle of about 140°. This arrangement allows for a direct σ-overlap between the d_{z^2} orbital of the iron atom and the $2p\pi_g$ orbital of the O_2 (its HOMO) which lies in the iron xz plane (Figure 37a). In addition the other $2p\pi_g$ orbital of the O_2 bonds *weakly* in a π-bonding fashion with the d_{yz} orbital of the iron (Figure 37b).

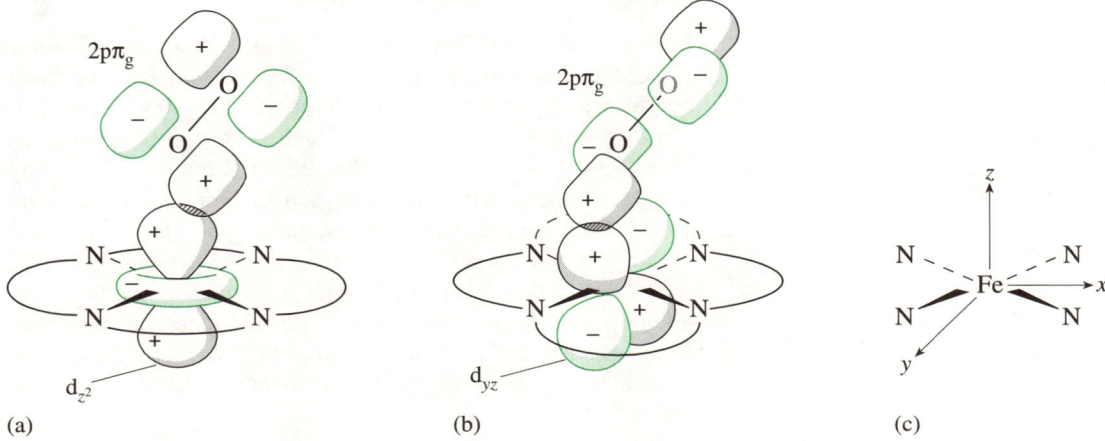

Figure 37 Overlap of O_2 and iron orbitals in oxymyoglobin. The $2p\pi_g$ orbitals of the oxygen molecule form (a) a σ bond with the d_{z^2} orbital on the iron, and (b) a weak π bond with the iron d_{yz} orbital; (c) the immediate environment of the iron atom, showing coordinate axes.

We are now in a position to sketch a simplified molecular orbital energy-level diagram for the Fe–O_2 complex. The molecular orbital diagram in Figure 38 shows how the energy levels change when the d_{z^2} orbital of the iron atom interacts in a σ-bonding fashion with one of the $2p\pi_g$ orbitals on the oxygen molecule, and how the other $2p\pi_g$ orbital interacts with the d_{yz} iron atomic orbital. Sigma bonding and antibonding orbitals are formed from the interaction of the iron d_{z^2} and one O_2 $2p\pi_g$ orbital. Pi bonding and antibonding orbitals are formed from the iron d_{yz} and the other O_2 $2p\pi_g$ orbital. The iron d_{xz} and d_{xy} orbitals do not overlap to any great extent with either O_2 $2p\pi_g$ orbital, and remain non-bonding. We have eight electrons to put into Figure 38, six from iron(II) and two from the $2p\pi_g$ orbitals of O_2 (see Figure 25).

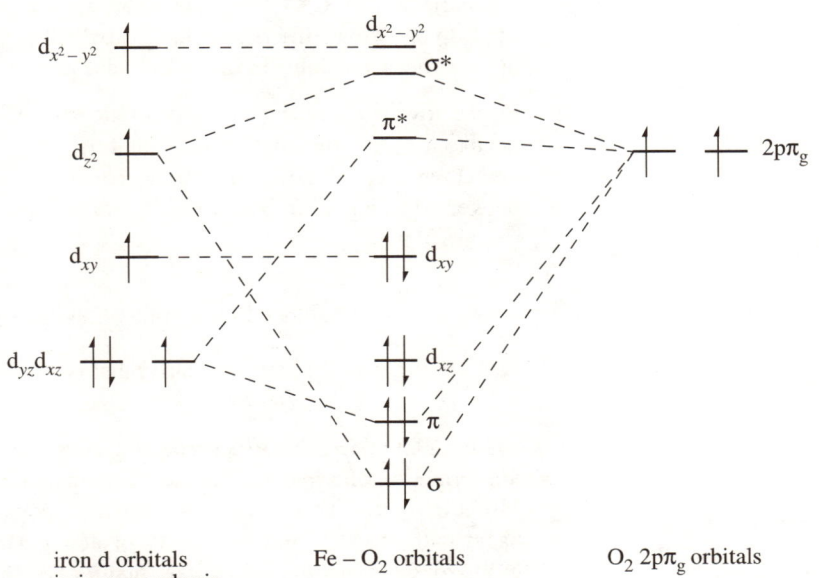

Figure 38 Molecular-orbital energy-level diagram for the Fe–O_2 complex in oxymyoglobin.

iron d orbitals in iron–porphyrin   Fe – O_2 orbitals O_2 $2p\pi_g$ orbitals

This is one of a few cases where it is difficult to assign unambiguously the oxidation state of either the iron or the O_2, since the relative energies of the O_2 orbitals and the iron orbitals before formation of the Fe–O_2 complex are very similar. The corresponding molecular orbitals that are formed in the Fe–O_2 complex have roughly 'equal' amounts of iron and O_2 character. In other words, the electrons in the molecular orbitals are shared roughly equally between iron and O_2. It is therefore not strictly correct to describe it as an Fe^{III}–O_2^- complex, although it can be said to contain some Fe^{III}–O_2^- character, along with some Fe^{II}–O_2 character. Another thing to notice about the diagram is that the electrons are all paired, and hence the complex is diamagnetic.

SAQ 10 Assuming that an Fe^{III}–O_2^- complex is formed, what would you expect the vibrational frequency of the bound O_2 to be? Explain your answer with reference to Figure 26.

There's one more point we should note about the arrangement of electrons in deoxy-Mb and oxy-Mb. In the deoxy state the iron(II) is clearly high spin. However, from the diamagnetism of the oxy state, the iron here must be low spin. Therefore, on O_2-binding, the iron goes from high spin to low spin.

Before we leave Mb to concentrate on the intricacies of Hb, it is worth examining the role of the distal histidine in Mb. The distal histidine is essential for the successful function of Mb. Without it, the protein has a much lower affinity for oxygen and a much higher affinity for other small molecules. This shouldn't be too surprising, because we have already seen in Figure 35 that the N—H group of the distal histidine forms a stable hydrogen bond with the further oxygen atom of the O_2 ligand, which increases the affinity of Mb for oxygen. Nevertheless, other small ligands will bind to the iron, essentially blocking the O_2 binding site. If this binding of small ligands is irreversible, the Mb ceases to function, with catastrophic consequences for the organism. Two examples of small ligands that are efficient at binding to iron are carbon monoxide and cyanide. Both substances are well known for their high toxicity to nearly all life forms. Indeed, both molecules are capable of irreversibly binding to Mb and blocking the access of O_2 to this binding site.

As the binding of both CO and CN^- to haem is so efficient, only very small concentrations of either are needed to block completely the O_2 binding site in Mb and Hb.

☐ What is the distinctive feature of the bonding of CO to transition metals that gives such strong bonding?

■ The binding stability of CO to a transition metal is enhanced via π-acceptor back-bonding (refer to Block 4, if necessary, to remind yourself).

Fortunately, the concentration of CN^- in the air is too low to cause any problems. By comparison with CN^-, the concentration of CO is relatively quite high, especially in built-up city areas. If gas fires have blocked flues, high concentrations of CO can build up, creating a poisoning hazard. CO is also present in vehicle exhausts and tobacco smoke.

To see how Mb prevents the catastrophic irreversible binding of CO to the haem site, we need to look carefully at the position of the distal histidine. Its position is such that it lies above the vacant binding site on the iron atom. As O_2 binds with an Fe—O—O angle of between 140 and 175°, the distal histidine does not sterically clash with the bound O_2 (Figure 39a).

☐ At what angle does CO normally bond to a metal?

■ CO ligands are bound linearly; that is, the Fe—C≡O angle is 180°.

In Mb, if CO does bind to the iron site, steric crowding by the distal histidine precludes the formation of a linear —Fe—C≡O complex (Figure 39b). In this way, CO binding to Mb is inhibited. This can be confirmed by comparing the CO binding affinity of free haem with the CO binding affinity of deoxy-Mb: the binding affinity of a free haem group for CO is about 25 000 times that for O_2 (CO is a much better ligand), whereas the binding affinity of Mb for CO is only about 250 times greater than that for O_2. In other words, CO is a very effective poison of the haem group, and, in the absence of the distal histidine, CO would bind so strongly to Mb that atmospheric O_2 could not displace it to any great extent. However, the presence of the distal histidine reduces the affinity of Mb for CO by a factor of 100 compared with free haem, and this is sufficient for the relatively high pressure of O_2 in the atmosphere to displace any CO bound to Mb.

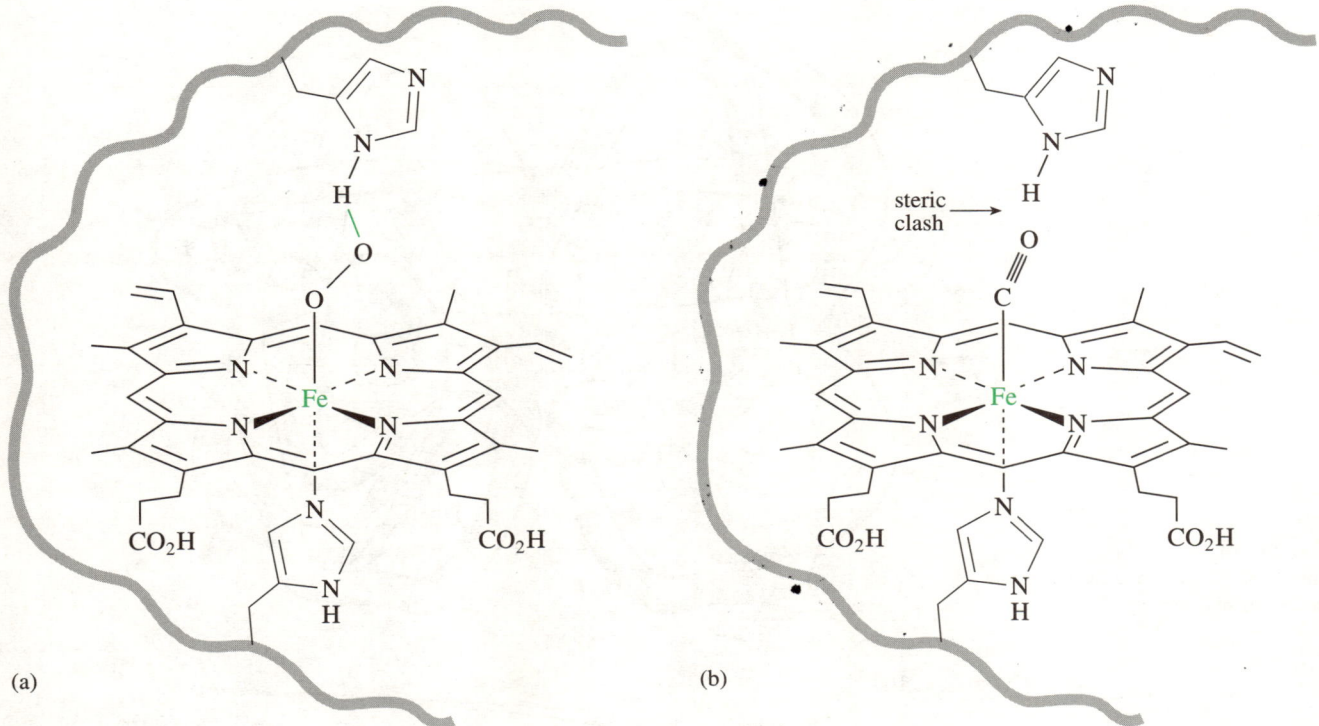

Figure 39 Active site structures of (a) oxymyoglobin and (b) carboxymyoglobin. In carboxymyoglobin the carbon monoxide sterically clashes with the distal histidine. (The Fe—C≡O angle is a matter of debate, but is thought to be in the range 140–175°.)

Remember that the CO partial pressure in air is relatively low, whereas that of O_2 is approximately 0.2 atm. Increasing the CO pressure in air eventually leads to effective competition of CO for the Mb binding sites: this is when CO poisoning can occur.

Haemoglobin and myoglobin have very similar CO binding characteristics (see Section 5.3.4). The toxicity of CO for both proteins is apparent from the following data: at 0.02 per cent by volume of CO in the atmosphere, 10 per cent of the haemoglobin is converted to carboxyhaemoglobin, producing shortness of breath on exertion. When the atmospheric concentration of CO reaches 0.2 per cent (which may arise from poorly ventilated gas fires or trapped car exhausts), the concentration of carboxyhaemoglobin reaches 80 per cent and rapid death ensues. The appearance of the blood is not a good guide to the cause of death, because carboxymyoglobin has a similar bright red colour to that of normal oxygenated blood.

Summary of Section 5.3.3
In Mb we have seen some general features of O_2-binding. These are:

- O_2 binds to iron in a haem group;
- O_2 binds in an end-on fashion at an angle of 140–175°;
- bound O_2 has a vibrational frequency less than free O_2, indicating formation of a reduced form of O_2 (i.e. O_2^-) in Mb;
- on O_2-binding in Mb, the iron coordination geometry changes from square pyramidal to near octahedral, and the iron atom moves into the plane of the porphyrin ring;
- the bound O_2 is stabilised by hydrogen-bonding to the distal histidine.

Having studied the structure and function of Mb, we can now turn our attention to the more complex Hb.

5.3.4 Haemoglobin

At first sight the structure of Hb appears to be very complex (Figure 40). However, it is easier to consider Hb as essentially four Mb units linked together. Therefore, the Hb molecule has four haem units and, hence, four O_2-binding sites. The active sites of the haem units in Hb are almost identical to the active site structure in Mb, complete with proximal and distal histidines. As a result, we might expect to see exactly the same O_2-binding properties in Hb and Mb. But this causes a problem. Remembering that Mb appears in muscle and Hb appears in the bloodstream, we can see that Hb must deliver O_2 to Mb. If both Hb and Mb had precisely the same O_2-binding characteristics, the

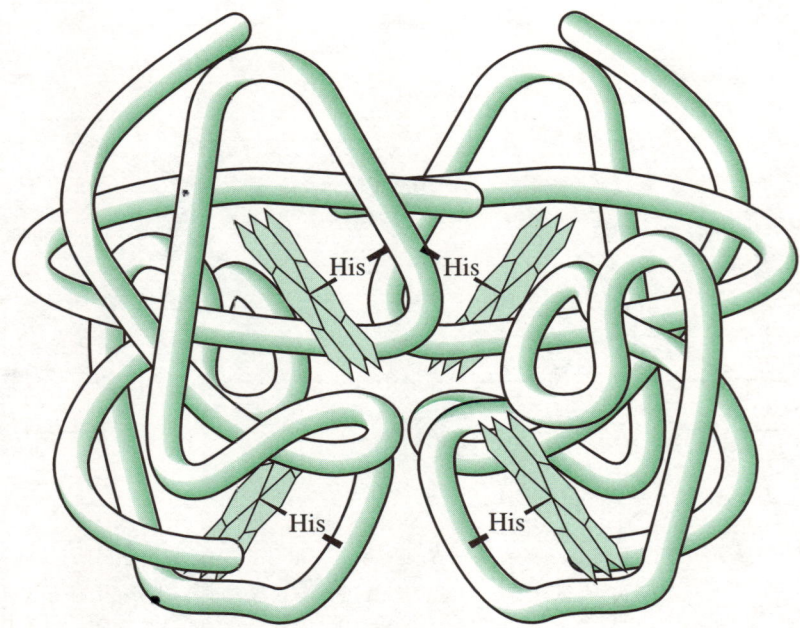

Figure 40 Higher-order structure of haemoglobin. The structure can be thought of as a myoglobin tetramer.

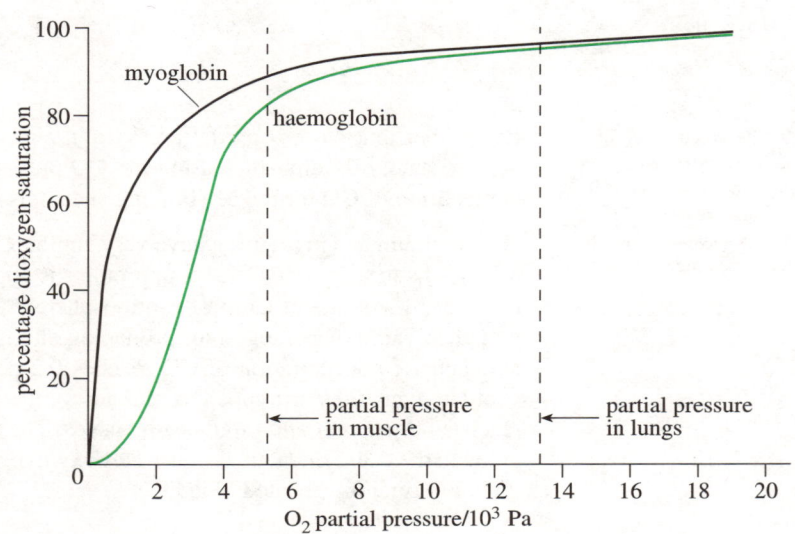

Figure 41 Percentage saturation of haem sites as a function of O_2 partial pressure for myoglobin and haemoglobin. Myoglobin follows a hyperbolic curve, whereas haemoglobin follows a sigmoidal curve due to a positive cooperative effect between the individual haem sites.

transfer of O_2 from Hb to Mb would be inefficient, certainly not efficient enough to maintain an appropriate O_2 concentration in the muscle tissue. Nevertheless, if we look at Figure 41, we can see that Mb *does* bind O_2 more efficiently than Hb at low partial pressures of O_2 (similar to the O_2 partial pressures found in a working muscle). What property of Hb causes this difference in O_2-binding efficiency, despite the structural similarity of the O_2-binding sites? To answer this question, we need to look at the structure of Hb in more detail.

☐ When O_2 bonds to Mb, how does the position of the iron change?

■ The iron moves into the plane of the porphyrin ring.

As in the Mb case, O_2-binding to the haem site in Hb causes a movement of the iron atom into the plane of the ring (Figures 34b and 35b). From very extensive structural studies of Hb, it turns out that this motion of the iron atom on O_2-binding causes changes in the rest of the protein's higher-order structure. These structural changes alter the O_2 affinity of the other haem sites in Hb. In other words the four haem sites in Hb can 'communicate' with one another, the communication being in the form of a structural change. This type of communication within a protein is known as an **allosteric interaction**. In the case of Hb, the binding of O_2 to one haem site increases the O_2 affinity of the remaining haem sites; this is known as the **cooperative effect**. (As the binding of the first O_2 *increases* O_2 affinity at the remaining sites, this is called a **positive cooperative effect**.)

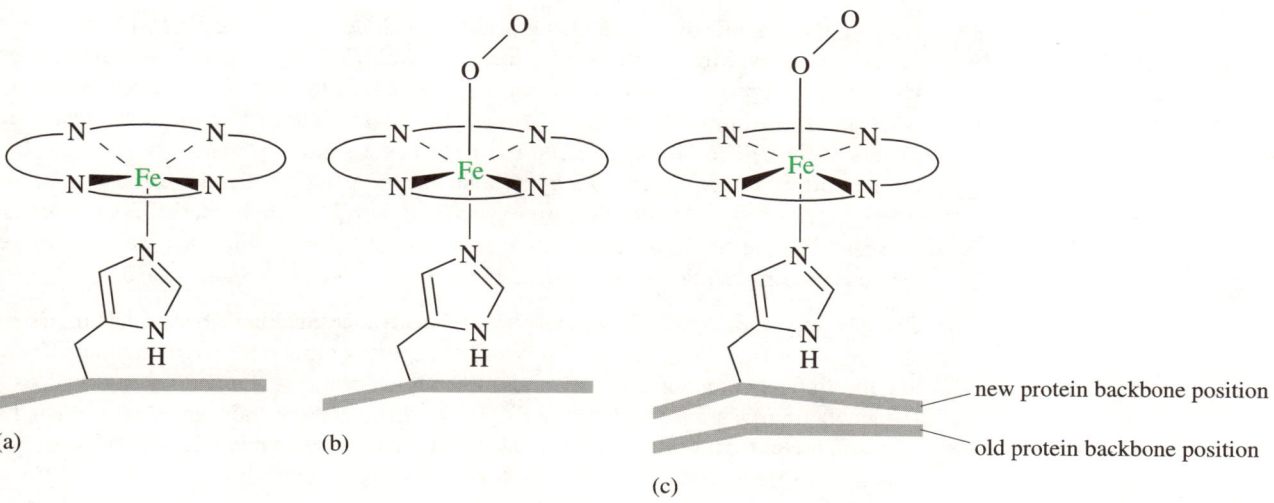

Figure 42 Haem group geometry changes on O$_2$-binding in haemoglobin; the haem group is abbreviated to four nitrogen atoms linked by curved lines: (a) deoxy active site showing iron atom out of porphyrin plane; (b) T-state, showing O$_2$ bound to iron atom, but iron atom is not fully in plane of porphyrin ring; (c) R-state, showing iron in plane of porphyrin ring and the motion of the protein backbone responding to the motion of the iron atom.

From a structural point of view, the changes in Hb on progressive O$_2$-binding are extremely subtle and still not fully understood. However, several key features have been observed in the crystal structures of the deoxygenated and the oxygenated forms of Hb. It appears that there are two distinct forms of Hb; these are known as the **tense form** (abbreviated to **T-state**) and the **relaxed form** (abbreviated to **R-state**). The active site of the T-state is shown in Figure 42. In completely deoxygenated Hb, the iron atom is out of the plane of the porphyrin ring in all the haem binding sites (Figure 42a). On the binding of one O$_2$ atom to the first haem site, the iron atom moves slightly towards the plane of the porphyrin ring, but not fully into the plane (Figure 42b). The reasons for this are fairly complex, but the proximal histidine is the root cause. It prevents the complete movement of the iron(III) into the plane of the porphyrin ring because it is anchored to the protein. The most stable iron(III) coordination geometry, for the singly oxygenated protein, is when the iron is in the plane of the porphyrin ring. But since the iron is slightly out of the plane of the porphyrin ring, the system is under tension — hence the term 'T-state'. At some point, probably on binding of further O$_2$ molecules to other haem sites, the tension becomes too great for the protein structure. At this point the iron(III) atoms move completely into the plane of the porphyrin rings (Figure 42c). This motion causes a significant change in the position of the proximal histidine, which, in turn, changes the position of the protein backbone; the protein structure is now in its relaxed state or R-state. But this is not the end of the protein movements. The change in the position of the protein backbone is mechanically transmitted to all of the haem sites. In the R-state, O$_2$-binding to deoxyhaem gives the iron(III)–O$_2$ complex, in which the iron atom is directly in the plane of the haem. Because the tension of the protein is released on transition to the R-state (after O$_2$-binding), this process has been called the **Perutz trigger mechanism** ('trigger' due to the nature of the process).

Perhaps the simplest way of depicting the trigger mechanism is shown in Figure 43. The two forms of the protein are shown. The R-state is favoured when Hb is oxygenated, the T-state is favoured on deoxygenation.

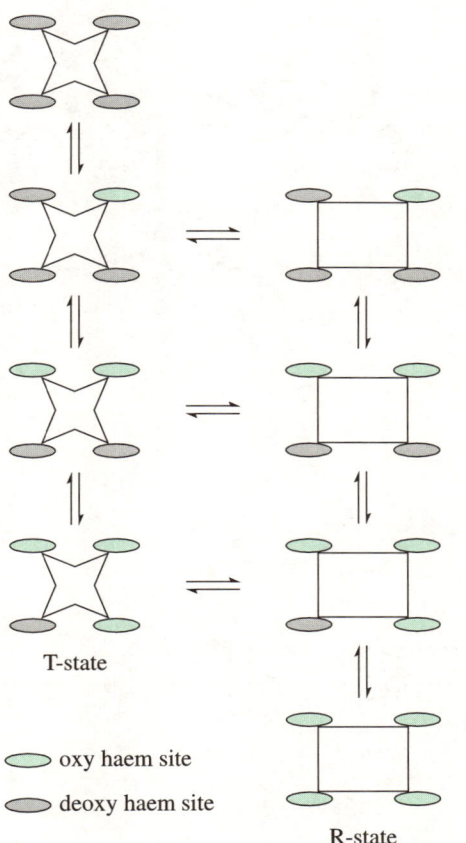

Figure 43 Schematic diagram of T- and R-states of Hb; the higher the degree of oxygenation, the more likely the R-state is to dominate.

How do these subtle structural changes affect the binding of O_2 to Hb? At low oxygen partial pressures, Mb can remove O_2 from Hb. At high oxygen partial pressures (when Hb must pick up O_2 as efficiently as possible — that is, in the lungs), the R-state dominates and the affinity of Hb for O_2 is relatively high. Figure 41 shows these results in graphical form. The O_2-binding curve for Hb has a sigmoidal shape. In contrast, O_2-binding affinity to Mb follows a hyperbolic shape. A hyperbolic curve shows that the percentage of Mb molecules that are oxygenated depends simply on the oxygen partial pressure. For the sigmoidal curve, the percentage of Hb molecules that are oxygenated depends not only on the oxygen partial pressure, but also on the state of Hb.

The Hb story does not end here. Despite all the advances that have been made in understanding how Hb works, the protein is still the centre of much research attention. There are questions that remain partially or completely unanswered. For instance, how does a foetus obtain oxygen from its mother's Hb? In order that the transfer of oxygen can be effected, foetal Hb must have a higher affinity for oxygen than maternal Hb. What are the structural differences between the two forms of Hb that cause this?

The structures and chemistry of both Mb and Hb are complex and yet fascinating. The haem groups and protein structure are linked perfectly to give the optimum O_2-binding, transport and storage system. Indeed, this is clearly seen if one takes a simple haem group (in the absence of any protein) and observes its O_2-binding properties. It is a well-known chemical fact that a free haem group reacts *irreversibly* with O_2 and could never act as an O_2-carrier on its own: this is because it reacts with oxygen in such a way that the O_2 molecule is eventually dissociated. The general reaction scheme for a model compound involving pyridine rather than histidine is shown in Figure 44. Two haem groups

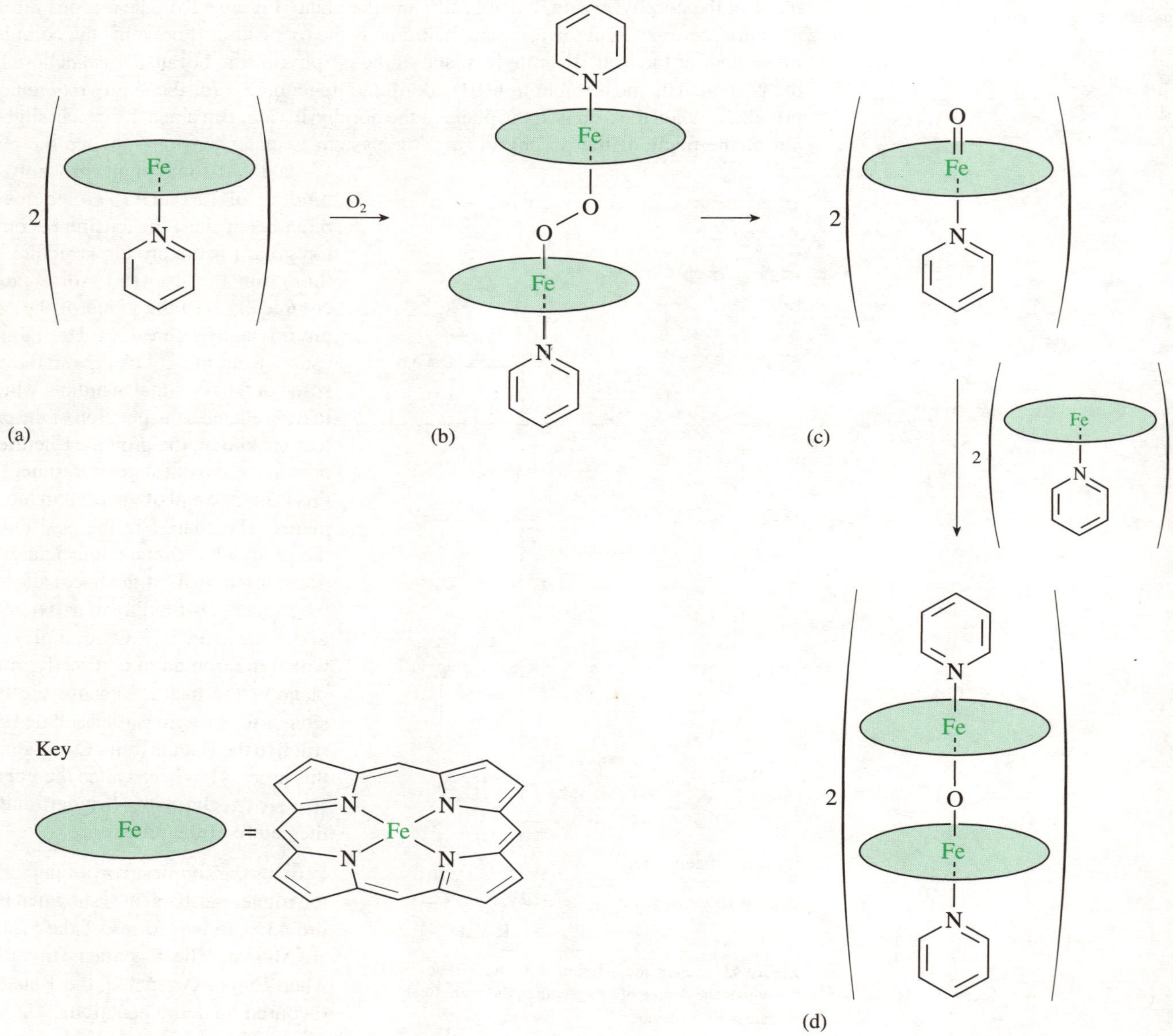

Figure 44 Reaction of free haem (with pyridine) group with O_2. In the first instance, a μ-peroxo haem dimer (b) is formed, which then cleaves to give an oxoiron(IV) haem complex (c). The oxo complex quickly reacts with further haem groups to give μ-oxoiron(III) haem dimers (d). The result is that the O_2 molecule has been irreversibly cleaved.

(Figure 44a) react with O_2 to give a μ-peroxo dimer (Figure 44b); this is analogous to the μ-peroxo dimer shown in reaction 15.

☐ Why does this dimerisation not take place in Mb or Hb?

■ The bonded O_2 is stabilised by hydrogen-bonding to the distal histidine. The steric requirements of the proximal and distal histidines and the protein backbone prevent the approach of a second haem group.

This dimer then dissociates to give the reactive iron(IV)–oxo haem (Figure 44c), which reacts with further haem groups to give μ-oxoiron(III)–haem dimers (Figure 44d) The overall result is that the oxygen molecule has been irreversibly split.

For quite some time it seemed impossible that an isolated haem group could recreate the reversible O_2-binding seen in Hb/Mb. However, in 1978, in a piece of simple but highly innovative thinking, James Collman hit on the idea that if the dimerisation of haem in the presence of O_2 could be prevented, then reversible O_2-binding might be observed.

☐ How might the dimerisation of the haem be prevented?

■ As in the natural systems, it might be possible to sterically prevent the approach of a second haem by putting other groups in the way.

Collman prepared the haem group derivative shown in Figure 45. He called the molecule an **iron picket-fence porphyrin**[G]. The haem group has four bulky groups attached to it. These groups all stick up around the O_2-binding site like a fence. This fence prevents two haem groups from coming together (but does not hinder the binding of O_2), and thus precludes the formation of μ-peroxy haem dimers following the binding of O_2. Indeed, Collman found that the iron picket-fence porphyrin could reversibly bind O_2 in an analogous fashion to the haem groups in Mb, and that the O_2 affinity of iron-picket fence porphyrin was very similar to that of Mb.

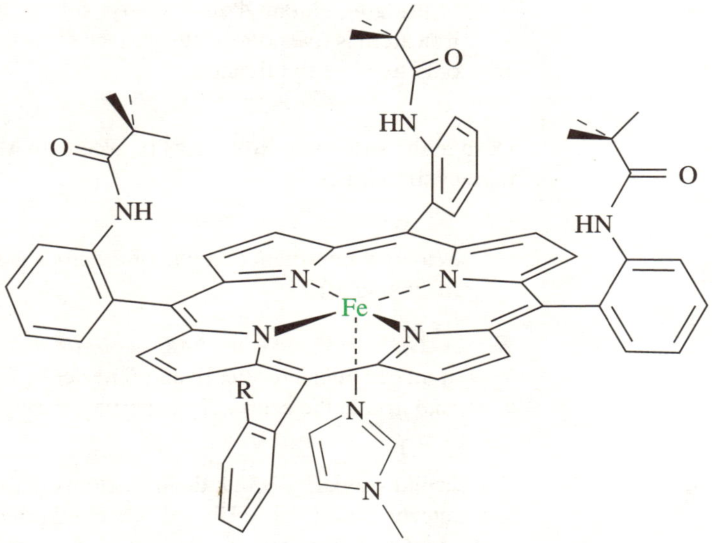

Figure 45 Collman's iron picket-fence porphyrin, showing a haem group, to which four aromatic groups are attached (between two adjacent pyrrole rings); the benzene rings have bulky side-chains in *ortho* positions (for clarity, one has been represented as 'R', and the haem side-chains have been omitted). The side-chains form a sterically crowded 'fence' around the site where O_2 binds to iron. This complex exhibits reversible O_2 binding, similar to that exhibited by myoglobin and haemoglobin.

The structures of Mb (Figure 31) and Hb (Figure 40) show that here too the haem groups are buried within the structures of the proteins; as with Collman's iron picket-fence porphyrin, this prevents dimerisation of the haem groups in the presence of oxygen. The burying of the oxygen binding site within the protein also serves another purpose. The bound oxygen is essentially a superoxide anion radical, which, if it were liberated into the cell, could irreversibly damage other biomolecules. By keeping the superoxide in a protective pocket, both Mb and Hb prevent this detrimental reaction. The other O_2-carrying proteins we shall study also prevent the reaction of bound O_2 by keeping it in a deeply buried protein pocket.

5.3.5 Haemerythrin and haemocyanin — two different O_2-carriers

We have already seen that some lower animals have different coloured blood from the brilliant red colour of blood that contains Hb. These different colours are qualitative evidence that the O_2-carrying protein in these lower animals is not Hb. Some invertebrates have blood that is colourless in the absence of oxygen yet a deep blue colour in the presence of oxygen (true blue blood!); the blood of certain marine worms is colourless in the absence of oxygen and deep burgundy in the presence of oxygen. However, the fact that all these species have coloured blood suggests that their oxygen-transport systems could involve transition-metal containing proteins. This should not be surprising because many transition metals other than iron can exist in a variety of oxidation states, and this is the key property for metals in O_2-carrying proteins. In this Section, we shall consider two other O_2-carrying proteins, **haemerythrin (Hr)** and **haemocyanin (Hc)**. It will be useful to compare and contrast the properties of these two proteins with what we already know about Hb and Mb.

(a) Haemerythrin

Haemerythrin[G] (Hr) occurs in the bloodstream of certain marine worms. It is a small protein with a relative molecular mass of about 13 500. Its higher-order structure has similarities to Hb (which we can consider to be a Mb tetramer), in that Hr is often found as an *octameric* unit, with a total relative molecular mass of about $8 \times 13\,500 = 108\,000$. At its active site, Hr contains two iron atoms. However, despite the 'haem' in its name, the protein does *not* contain a haem group. From single-crystal X-ray diffraction studies the structure of the active site in the deoxy form has been determined; it is shown in Figure 46.

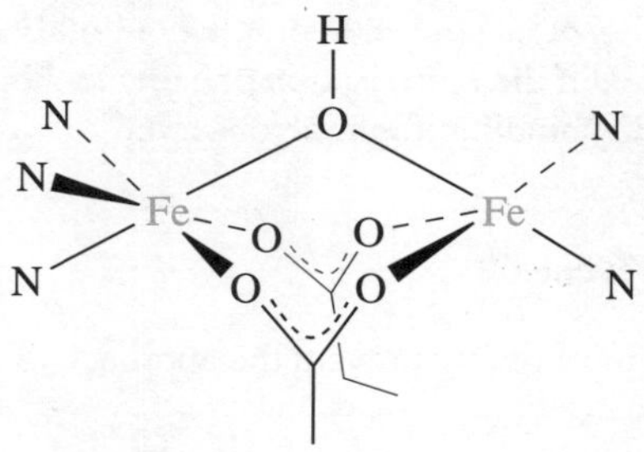

Figure 46 Active site of deoxyhaemerythrin. Each iron atom is coordinated either by two or three histidine amino acid side-chains (indicated by 'N'); they are also bridged by the carboxylate groups of glutamate and aspartate amino acid side-chains. A hydroxide group also bridges between the two iron sites.

The first point to make about the active site structure of deoxy-Hr is that it is completely different from that of Hb and Mb. In deoxy-Hr there are two iron atoms coordinated by several amino acid side-chains and a hydroxide group; the aspartate and glutamate side-chains and the hydroxide are bridging ligands.

☐ What is the difference in coordination between the two iron atoms?

■ One of the iron atoms is six coordinate in a roughly octahedral geometry. Its coordination sphere is made up of the nitrogens of three histidyl side-chains and the oxygens of bridging glutamyl and aspartyl side-chains and a hydroxide anion. The second iron atom is five coordinate with the same three bridging ligands as the first iron but only two histidyl ligands.

Despite the structural differences between Hb/Mb and Hr, there is a couple of very important similarities.

☐ Why is it important that one of the irons is five coordinate? Where does this occur in Hb and Mb?

■ A five-coordinate iron atom is *coordinatively unsaturated*. In other words this iron atom can bind a further ligand. The deoxy forms of Hb and Mb contain five-coordinate iron in the centre of the porphyrin ring.

The second similarity is that the iron atoms in the deoxy form are both in the +2 oxidation state, again similar to Hb and Mb. Our knowledge of how O_2 binds to the active site of deoxy-Hr comes from single-crystal X-ray diffraction of several Hr derivatives and from resonance Raman studies.

The resonance Raman spectroscopy studies proved to be particularly revealing. For resonance Raman spectroscopy, more information can be obtained using ^{18}O-labelled O_2 as an oxygen source.

☐ What is the most abundant isotope of oxygen?

■ ^{16}O; this occurs naturally in 99.76 per cent abundance, whereas the natural abundance of ^{18}O is 0.20 per cent.

The labelled O_2 gas contains not just the normal ^{16}O isotope but contains an increased proportion of ^{18}O atoms. The 50 per cent-labelled gas itself contains a statistical mixture of isomers, 25 per cent $^{16}O-^{16}O$, 50 per cent $^{18}O-^{16}O$ and 25 per cent $^{18}O-^{18}O$.

SAQ 11 (a) What experimental technique can be used to obtain the vibrational spectrum of O_2?

(b) How many peaks are expected in the vibrational spectrum of 50 per cent ^{18}O-labelled O_2? Explain your reasoning.

(c) Which peak relates to which isomer?

Figure 47 depicts the O_2 part of the resonance Raman spectrum ot oxy-Hr oxygenated with 50 per cent ^{18}O-labelled O_2. (In Section 4.3 we looked at how resonance Raman spectroscopy worked, and what information it gave us about protein active sites.)

Figure 47 Resonance Raman spectrum of oxyhaemerythrin oxygenated with 50 per cent ^{18}O-labelled O_2. The wavelength of the excitation radiation used was 514.5 nm.

☐ What is immediately striking about the stretching frequencies in Figure 47?

■ The Raman frequencies due to the O—O bond are now much lower than in free molecular O_2 (Figure 26).

☐ What oxygen species would give rise to these frequencies?

■ A glance at Figure 26 shows that these frequencies correspond to a peroxide. In other words, the oxygen must have formally gained two electrons from the protein to give O_2^{2-}.

You may also have noticed that there are now *four* bands in the spectrum, corresponding to four vibrations, as opposed to three for free 50 per cent-^{18}O-labelled O_2. The two flanking bands correspond to $^{16}O-^{16}O$ and $^{18}O-^{18}O$ vibrations*. The central band due to the $^{16}O-^{18}O$ vibration can be considered as two vibrations.

How can we account for the central $^{16}O-^{18}O$ vibration splitting into two bands? There are two possible ways of bonding $^{16}O-^{18}O$ to a single five-coordinate iron atom: these are with either the ^{16}O bonded directly to the iron (structure **14**) or the ^{18}O bonded directly to the iron (structure **15**); see Figure 48. These two structures do have slightly different $^{16}O-^{18}O$ stretching frequencies, so we would expect to see four O—O vibrations in the resonance Raman spectrum.

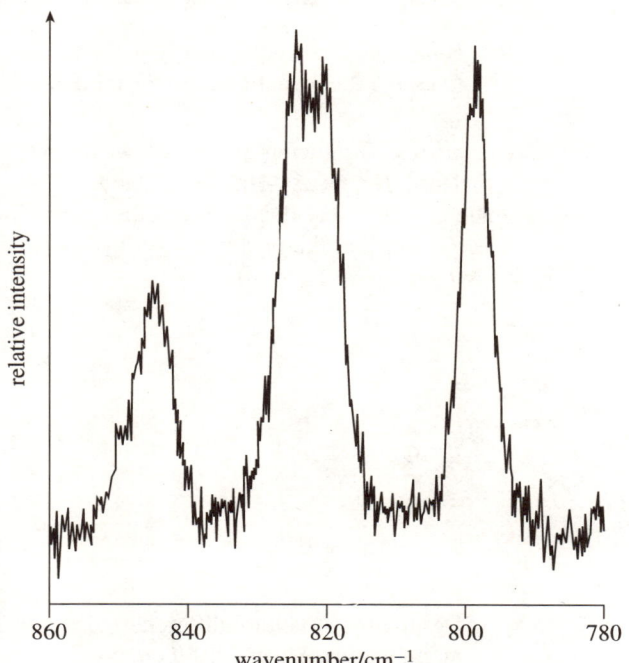

Figure 48 Two possible bonding geometries of $^{16}O-^{18}O$ to an iron atom.

* With equal proportions of Hr-$^{16}O-^{16}O$ and Hr-$^{18}O-^{18}O$, we would expect absorptions of similar peak height. The fact that the absorption at c. 795 cm^{-1} is larger than the one at c. 845 cm^{-1} may indicate that the isomeric ratio was not 1 : 1, or that another vibration also has an absorption at the lower frequency.

What are our conclusions from the resonance Raman experiments? The number of bands in the resonance Raman spectrum strongly suggest an η^1 coordination of O_2 to the active site. The vibrational frequency of the bound O_2 in the spectrum of oxy-Hr is very low and shows that it is bound as the peroxide form, so it must have received two electrons from the active site.

☐ Which iron atom do we expect the O_2 to bond to?

■ Remembering the structure of the deoxy form from Figure 46, it seems reasonable that in oxy-Hr, O_2 binds in an η^1 fashion to the five-coordinate iron atom.

☐ The evidence shows that the oxygen binds formally as O_2^{2-}. What will have happened to the oxidation state of the iron atoms?

■ Both iron atoms are formally oxidised to iron(III), but remember that it is difficult to assign the oxidation state unambiguously.

This process is shown in Figure 49, where the structure of the active site is indicated in both deoxy-Hr and oxy-Hr. The structure of the oxy-Hr active site has been inferred from single-crystal X-ray diffraction studies on various Hr derivatives. These crystal structures predict that the O_2 binds as a hydroperoxide (taking a hydrogen atom from the bridging hydroxide group).

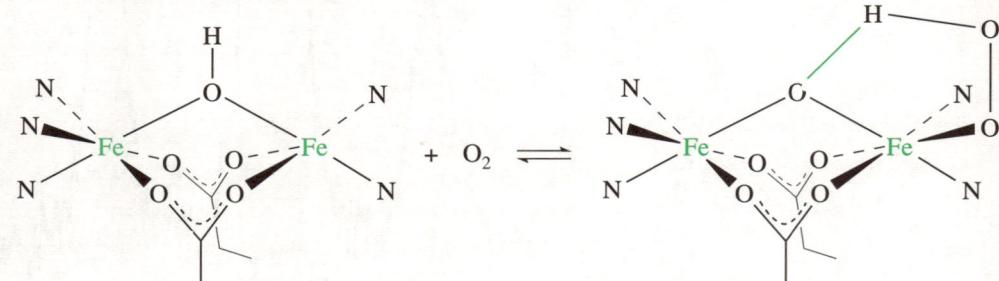

Figure 49 Binding of O_2 to haemerythrin and structure of the active sites. Note that there is an apparent transfer of hydrogen from the bridging hydroxide to the bound oxygen molecule; this is thought to help stabilise the binding of O_2 similar to the role of the distal histidine in myoglobin and haemoglobin. The resultant hydroperoxide is hydrogen-bonded to the oxide bridge.

☐ Despite the structural differences between Hr and Hb/Mb, there are several common features. Try and list them now.

■
- the reduction of O_2 to a lower oxidation state by iron;
- the formal oxidation of iron(II) to iron(III);
- η^1-binding of O_2 to a five-coordinate iron atom;
- the stabilisation of the bound O_2 with a hydrogen bond.

It is remarkable that such different proteins have so many common features.

(b) Haemocyanin

The third, and last, type of O_2-carrying protein we shall study is called haemocyanin[G] (Hc). This protein occurs in the bloodstreams of molluscs (e.g. snails) and arthropods (e.g. spiders). It is a very large protein with a relative molecular mass of around 1 000 000. In the deoxygenated state it is colourless, whereas in the presence of O_2 it is blue. At its active site, Hc has two *copper* atoms. In fact, the name 'haemocyanin' is a double misnomer, as the protein contains neither a porphyrin ring nor an iron atom. The structure of the active site is shown in Figure 50.

Structurally, the active site in Hc is similar to that in Hr, with two metal ions available for O_2-binding. The copper atoms in the deoxy form are in the +1 oxidation state.

☐ What is the electronic configuration of a copper(I) ion? Would you expect it to exhibit d–d transitions in a u.v./visible spectrum?

■ Copper(I) is d^{10}. This helps explain why the deoxy protein is colourless; as the d orbitals are full, there cannot be any d–d electronic transitions.

Copper(I) normally adopts a tetrahedral geometry; hence it has a coordination number of four. In this active site, each copper is coordinated by three nitrogen atoms of histidyl

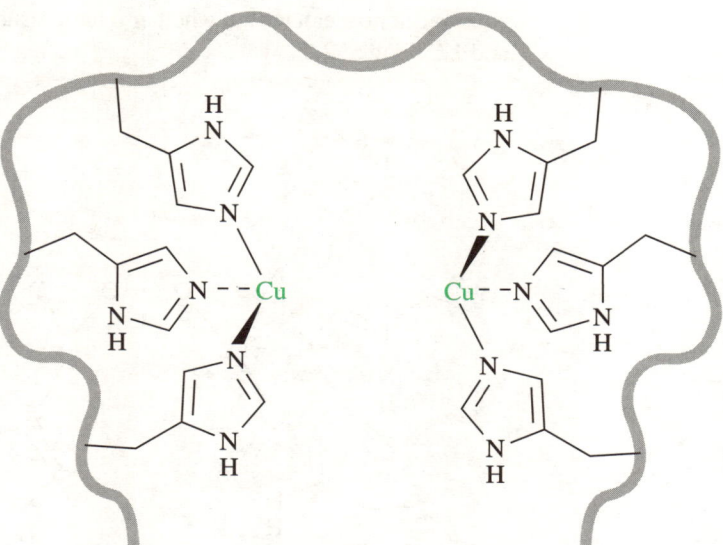

Figure 50 Active site of deoxyhaemocyanin. The copper-to-copper distance is about 360 pm (for scale, each copper-to-nitrogen bond length is around 190 pm).

side-chains. As in the previous cases of deoxy-Hb, deoxy-Mb and deoxy-Hr, the metal atoms are coordinatively unsaturated.

☐ What is different about this active site from the other three that we have studied?

■ In deoxy-Hc, *both* metal atoms are coordinatively unsaturated, so that both atoms can accept another ligand.

How does O_2 bind to the active site in the oxygenated form? The first piece of evidence in answering this question came, once again, from resonance Raman studies. In an exactly analogous manner to the experiment carried out with Hr, 50 per cent ^{18}O-labelled O_2 was used in the resonance Raman experiments. The results are shown in Figure 51.

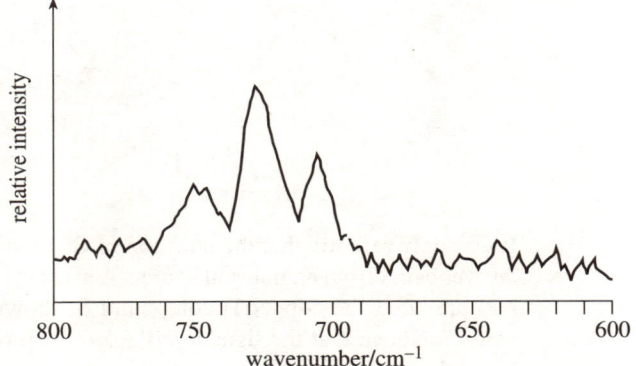

Figure 51 Resonance Raman spectrum of oxyhaemocyanin oxygenated with 50 per cent ^{18}O-labelled O_2.

☐ What are the two key features that you should look for in this spectrum?

■ (i) The position of the peak gives information on the strength of the O—O bond and therefore the extent of its reduction.

(ii) The number of peaks gives us a clue about the mode of bonding to the metals.

Firstly, we note that the O—O vibrational frequency is about 730 cm^{-1}.

SAQ 12 What does this frequency tell us about the charge on the bound O_2? What can we say about the formal oxidation state of the copper in the oxygenated form?

Secondly, the spectrum contains only three peaks. In contrast to Hr, the resonance Raman spectrum of Hc shows no broadening or splitting of the $^{16}O-^{18}O$ vibration. Hence the O_2 must be bound *symmetrically* at the active site. Reference to Figure 27 shows that we can rule out a structure of type **10** as a possible O_2-binding geometry. We can also rule out structure **11**, since it is very likely that the O_2 will bind to both copper ions simultaneously. But how can we tell whether it has a structure of type **12** or type **13** (structures **16** and **17**, Figure 52)?

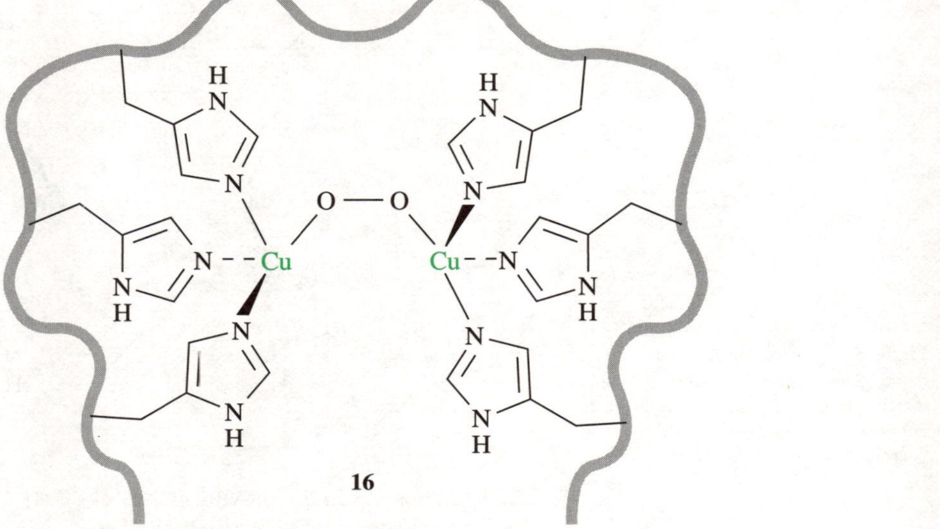

16

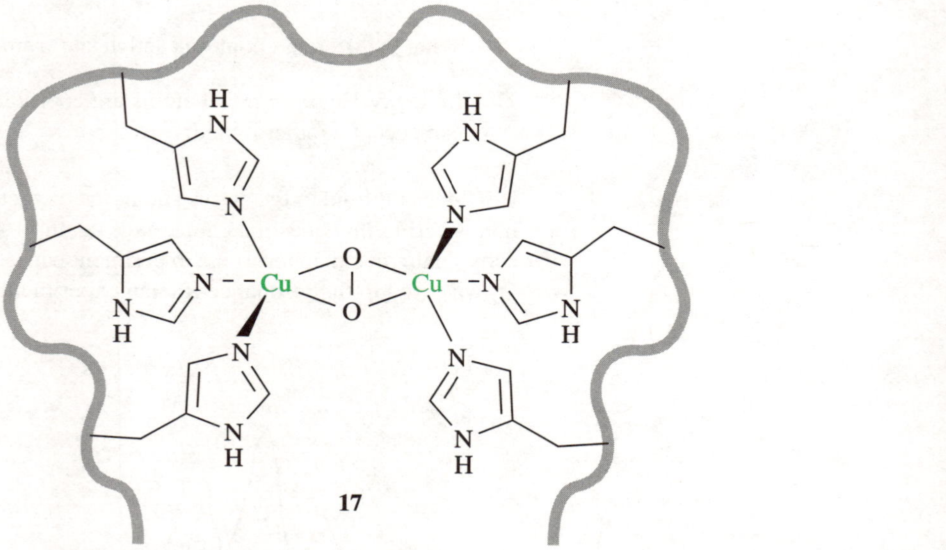

17

Figure 52 Two possible O$_2$ binding geometries in oxyhaemocyanin compatible with resonance Raman evidence. In both cases the O$_2$ is bound symmetrically between the two copper atoms.

It was only recently that the question of O$_2$-binding geometry in Hc was answered. The answer came from an unusual source. A group of Japanese chemists, led by N. Kitajima, had prepared the copper(I) compound **8**, shown on the left of reaction 15 (repeated below). Notice that the ligand, *tris*[3,5-di-isopropyl(pyrazoyl)borate] mimics the three histidyl side-chains coordinated to copper in Hc. The Japanese chemists noticed that their compound reversibly bound O$_2$ at low temperatures; the colour changes they saw matched exactly the colour changes seen on reversible oxygenation of Hc. They were able to obtain a single-crystal X-ray diffraction structure of the oxygenated product of their reaction; its structure (**9**) is shown as the product of reaction 15. Notice that on oxidation a dimer has formed.

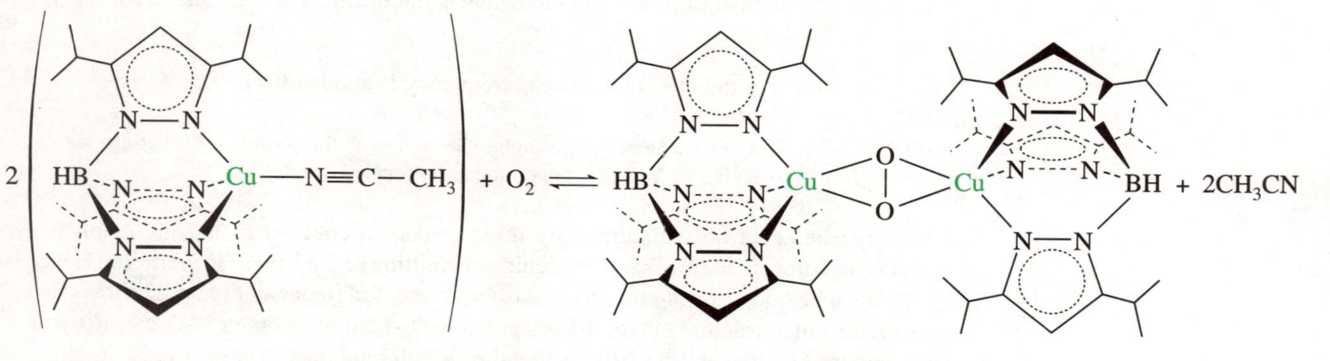

8   **9**   15

□ What is the nomenclature for the type of bridging formed by the O₂ here?

■ Oxygen has bound between the two coppers in a $\mu\text{-}\eta^2,\eta^2$ fashion.

Remarkably, the ultraviolet/visible spectrum of the oxygenated compound **9** matched almost *exactly* the ultraviolet/visible spectrum of oxy-Hc (Figure 53). This was good evidence that in oxy-Hc, O₂ binds between the two coppers as a $\mu\text{-}\eta^2,\eta^2$ peroxide. Indeed, this was confirmed by a single-crystal X-ray diffraction study of oxy-Hc in 1994.

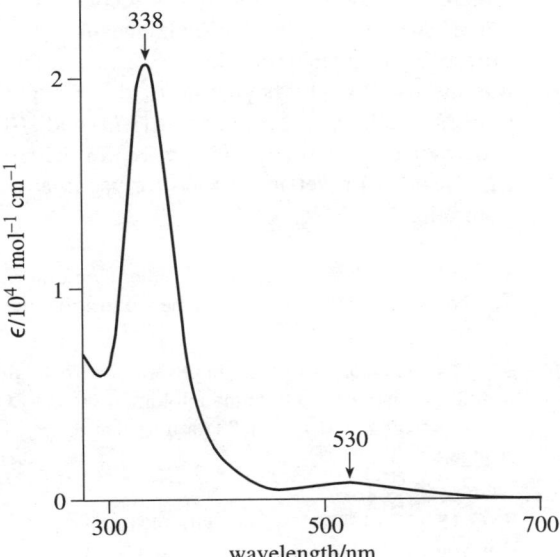

Figure 53 Ultraviolet/visible spectrum of compound **9**, with bands at 530 nm (ϵ = 840 l mol⁻¹ cm⁻¹) and 338 nm (ϵ = 20 800 l mol⁻¹ cm⁻¹). The u.v./vis. spectrum of oxyhaemocyanin has two similar bands at 570 nm (ϵ = c. 1 000 l mol⁻¹ cm⁻¹) and 345 nm (ϵ = c. 20 000 l mol⁻¹ cm⁻¹).

5.4 Summary of Section 5

We have seen three types of O₂-carrying groups that occur in biological systems; these are Mb/Hb, Hr and Hc.

1 Despite the structural differences between them, they all have the following key common features:

- the active site incorporates transition metal(s) that can change oxidation state;
- when oxygen is bound to a metal atom, the O—O bond is weakened;
- in the deoxy form at least one metal is coordinatively unsaturated;
- O₂-binding is accompanied by changes in the colour of the protein.

2 The active centre in myoglobin consists of a five-coordinate (square pyramidal) iron(II) atom. It is coordinated by the four nitrogens of a porphyrin ring and the proximal histidine. A more distant histidyl residue, known as the distal histidine, lies on the other side of the ring. In the deoxygenated form the iron atom lies 60 pm out of the plane of the ring, towards the proximal histidine.

3 In the oxygenated form of Mb, the O₂ binds in η^1 fashion to the iron, at an angle of 140–175°, and also forms a hydrogen bond to the distal histidine. In this form, the iron lies in the plane of the ring. The O₂ is reduced to superoxide, O_2^-, and iron is formally oxidised to iron(III).

4 The linear bonding of CO and CN⁻ to iron is destabilised by the presence of the distal histidine, which partially blocks the site.

5 Haemoglobin contains four haem sites. In its deoxygenated form it is in the tense or T-state. At high oxygen partial pressures, the oxy-Hb form is favoured and the iron atoms move into the planes of the rings, causing a movement of the protein backbone to the relaxed or R-state.

6 Haemerythrin has an active site containing two iron(II) atoms. In the deoxygenated form of Hr, the iron atoms are connected to three and two histidyl residues, respectively, and are bridged by three ligands — glutamate, aspartate and hydroxide.

7   On oxidation of Hr, oxygen bonds η^1 to the coordinatively unsaturated iron atom, and is reduced to peroxide, O_2^{2-}, whereas both iron(II) atoms are formally oxidised to iron(III).

8   Haemocyanin contains two copper(I) atoms in its active site, both coordinated to three histidyl side-chains. On oxidation, a symmetrical $\mu\text{-}\eta^2,\eta^2$ peroxo bridge is formed between the two copper atoms, which are both formally oxidised to copper(II).

Having seen how reversible O_2-binding is successfully carried out in proteins, another question comes to mind; how do proteins/enzymes irreversibly bind O_2? After all, the proteins we have discussed in this Section merely transport O_2 to where it is consumed. There must be some proteins/enzymes that are capable of taking the delivered O_2 and irreversibly using it in some biochemical process(es). Take respiration, for example, which was discussed at the beginning of this Section; how is the oxygen consumed in this process, and how is it converted to CO_2 and H_2O without releasing all the intermediate toxic molecules shown in Figure 24? In the next Section we shall look at some of the enzymes that irreversibly fix O_2 for use in a range of biochemical processes including respiration.

SAQ 13   Which transition metals, other than iron and copper, could act as reversible binders of oxygen? What features must these new transition metals exhibit?

SAQ 14   How does myoglobin prevent the irreversible binding of CO to the haem group? How would the relative affinity of myoglobin for O_2 and CO change if the distal histidine were changed to (a) tyrosine and (b) valine? (Assume that the rest of the protein structure is unaffected by the change.)

SAQ 15 $[Co(CN)_5]^{3-}$ reacts with O_2 to give a $[(CN)_5Co\text{—}O\text{—}O\text{—}Co(CN)_5]^{6-}$ dimer. Explain how you would determine the mode of O_2-binding.

SAQ 16   The affinity of myoglobin for CO is 250 times greater than for O_2. Why don't we all die of CO poisoning?

6 O_2-ACTIVATING AND H_2O_2-ACTIVATING ENZYMES

We have seen in Section 5 that haemoglobin transports O_2 around the body, and that O_2 is bound by myoglobin in muscle tissue. But where is the O_2 delivered to and how does it react? These are important questions to ask, since we know that O_2 is a very reactive molecule, capable of forming potentially harmful radicals (see Section 5, Figure 24). To avoid the production of these radicals, the reactions of O_2 within the body must be very carefully controlled. Not surprisingly, this control is exerted by several enzymes[*].

In this Section we shall briefly look at what is known about enzymes that catalyse reactions involving O_2, superoxide ($O_2^{\cdot-}$) and hydrogen peroxide (H_2O_2). These include the reaction of O_2 in aerobic respiration and the reaction of O_2 with biochemical molecules such as hormones, lipids and amino acids. We shall also study two classes of enzymes which catalyse reactions of H_2O_2. All these enzymes contain metals at their active sites.

6.1 O_2 in aerobic respiration — cytochrome c oxidase

We know from the equation for the aerobic respiration of glucose:

$$C_6H_{12}O_6(aq) + 6O_2(g) = 6CO_2(g) + 6H_2O(l) \qquad \textbf{10}$$

that oxygen must be consumed within our bodies in the respiration process. Obviously our bodies do not consume oxygen as in a flame, as this would liberate uncontrollable amounts of heat. Oxygen consumption in living systems is a very carefully controlled reaction, which has been referred to as *cold combustion*. The respiration sites in our

---

[*] An enzyme is a protein that *catalyses* a biochemical reaction, meaning that the enzyme is regenerated at the end of the reaction. Neither haemoglobin nor myoglobin are enzymes, because they do not catalyse an overall reaction.

bodies are the mitochondria^G, which are found within cells (see Figure 54 for a cross-section through a typical eukaryotic^G cell). The mitochondria contain a heavily folded inner membrane. The folds form pockets known as cristae^G (Figure 55). These pockets are the important sites for oxygen reaction in respiration. Oxygen is delivered into the cristae, where it is bound and reduced by one particular protein that is found embedded in the membrane. The protein is called **cytochrome c oxidase (CCO)**^G.

Cytochrome c oxidase is an enzyme, which, as its name implies, oxidises cytochrome c as part of its function. Cytochrome c is a smaller soluble protein (relative molecular mass c. 12 000), which transfers electrons from other parts of the mitochondrial membrane to CCO. Cytochrome c contains a haem group, in which the iron oxidation state changes between 2 and 3. The iron serves as the site in the protein where the electron is 'stored' during transport. As in the haem group in myoglobin or haemoglobin, the cytochrome iron is coordinated by a porphyrin ring and a histidyl group. However, unlike the oxygen-carrying proteins, the iron is also coordinated by a further amino acid side-chain (methionyl) on the opposite side of the porphyrin ring to the coordinated histidyl. Therefore, in cytochrome c the iron is six coordinate and coordinatively saturated. There are also other types of cytochromes in the mitochondrial membrane, and in other biochemical systems where electron transport is important. All of them are haemproteins^G containing coordinatively saturated iron.

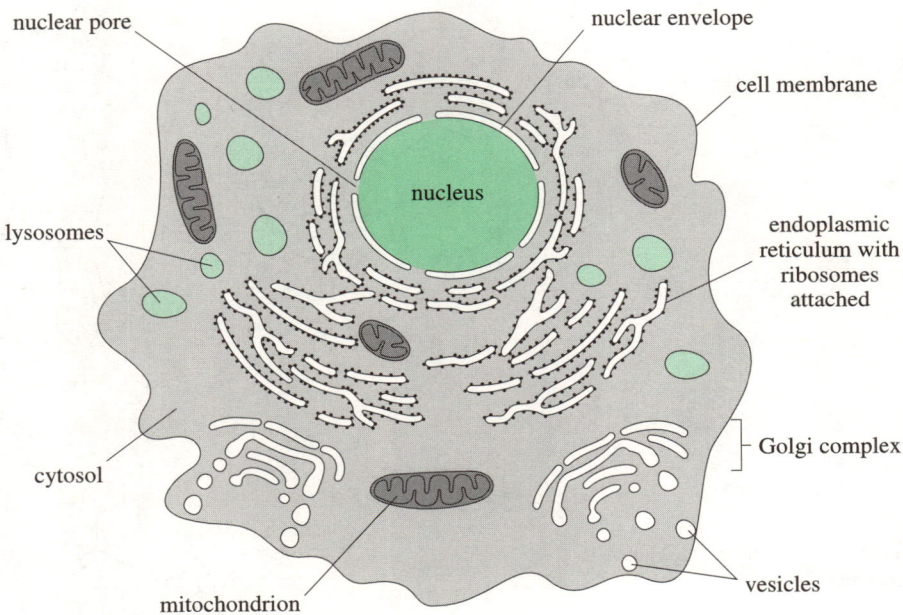

Figure 54 Schematic sketch of a eukaryotic cell, showing subcellular structures.

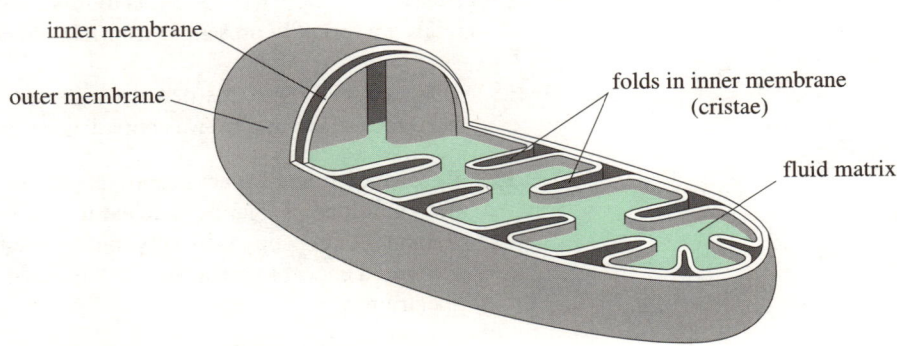

Figure 55 Schematic sketch of mitochondrial inner structure.

CCO is a very large and complicated protein (relative molecular mass > 100 000). Its overall, gross structure is known, but, as no crystal structure of the whole protein exists, the finer details of the structure are still unclear. (A crystal structure has been obtained for part of the protein.) It is known that CCO contains a variety of metal active sites, which are all important in the protein's mechanism of action. We shall concentrate on one site in particular, the site where O_2 is bound and reduced to water in the following reaction:

$$O_2(g) + 4H^+(aq) + 4e^- = 2H_2O(l) \qquad 16$$

The O_2-binding site contains two metal atoms, iron and copper. The iron atom is coordinated by a porphyrin ring. A coordination bond between a histidyl side-chain and the iron atom (cf. haemoglobin and myoglobin) anchors the haem group to the rest of the protein. The copper atom is coordinated by three histidyl side-chains, and is on the opposite side of the haem group to the histidine (Figure 56). The active site itself is reminiscent of the active sites in both myoglobin and haemocyanin.

☐ In which other example did we encounter two different metals at the active site?

■ The first example was the enzyme superoxide dismutase, which contains copper and zinc.

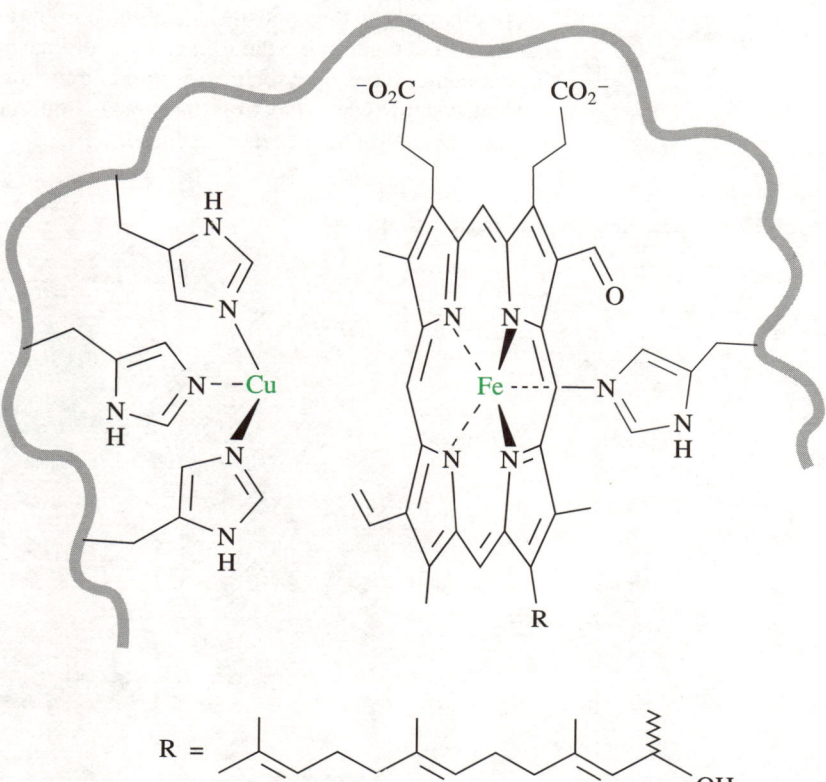

Figure 56 Active site for O_2-binding in cytochrome c oxidase.

The probable processes of O_2-binding and reduction are shown in Figure 57. In the reduced, active form of CCO (structure **18**, Figure 57) the O_2 binding site contains high-spin iron(II) and copper(I). O_2 binds to this form to give a peroxide (μ-η^1,η^1), in which the metals are formally present as iron(III) and copper(II) (structure **19**, Figure 57).

☐ If we could isolate this peroxide species, what experimental technique could be used to confirm that O_2 was bound as peroxide?

■ Resonance Raman spectroscopy is probably the best technique. This would show the vibrations of ligands attached to the iron or copper (remember that resonance Raman spectroscopy is usually metal-specific). Assuming that O_2 is bound as O_2^{2-}, we would expect to see a band in the region 750–920 cm^{-1} of the resonance Raman spectrum.

It is believed that the peroxide is quickly converted to hydroperoxide by the addition of a proton; at the same time, the iron(III) is reduced back to iron(II) (structure **20**, Figure 57). Addition of a further proton leads to the heterolytic cleavage of the O—O bond to give a H_2O molecule (coordinated to the copper) and an oxygen atom, which is believed to form a double bond with the iron atom, to give an unusual porphyrin–iron(IV)–oxo species (structure **21**, Figure 57). The iron(IV)–oxo ion is very unstable, and quickly reacts with

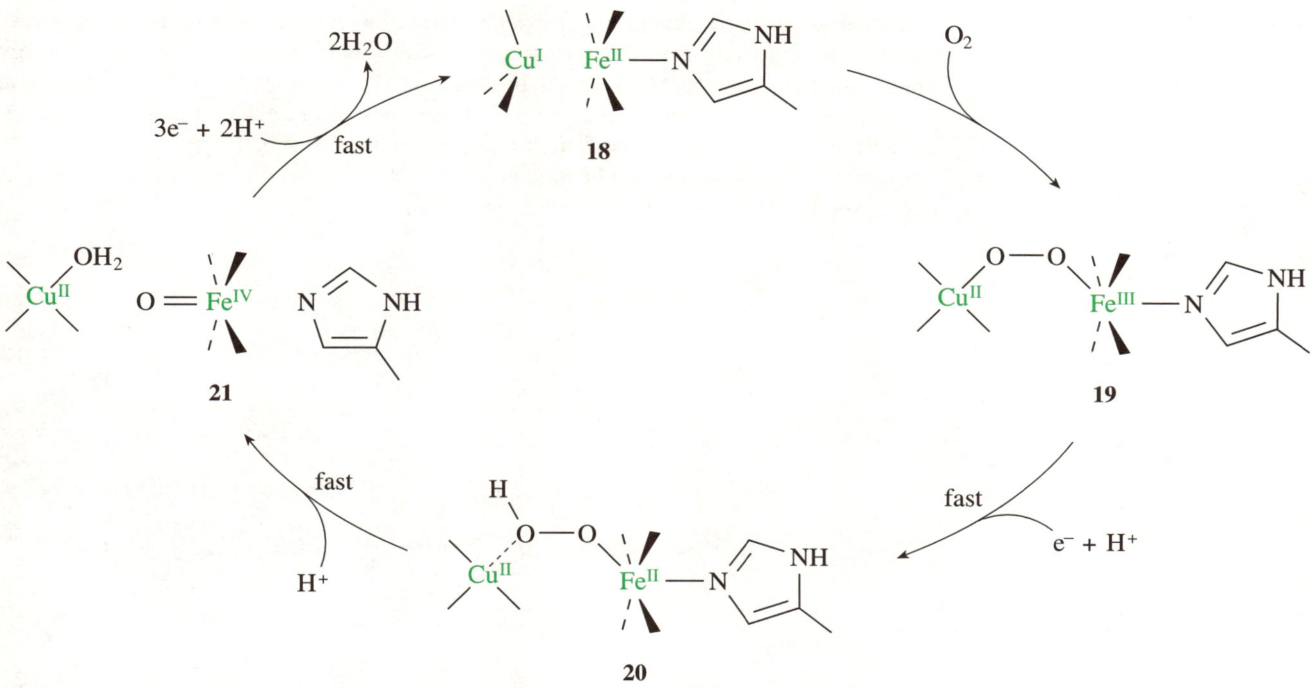

Figure 57 Catalytic cycle for the reduction of O_2 by cytochrome c oxidase.

more protons and electrons; a molecule of water is lost, leaving the starting protein, **18**.

After the formation of the oxygenated species **19**, the other steps in the catalytic cycle are all fast, which means that the intermediate peroxide, hydroperoxide and iron(IV)–oxo species cannot escape easily from CCO. In this way, partially reduced O_2 molecules and radicals are not released into the cell in large quantities. The electrons that are supplied to the protein come originally from glucose catabolism[G] (or other energy sources). As such, O_2 reduction at CCO represents the last step in the *aerobic catabolism* of glucose.

SFC 4

Most of the O_2 that is used for respiration is completely reduced to H_2O. Despite the high efficiency of CCO, however, some partially reduced O_2 molecules do escape into the rest of the cell. These potentially harmful molecules must be destroyed by the cell. This is accomplished by enzymes that act as detoxifying agents. Two particular classes of enzymes, peroxidases[G] and catalases[G], specifically decompose H_2O_2. One other class, called superoxide dismutases, catalyses the decomposition of the superoxide radical anion, $O_2^{-\cdot}$. We shall examine examples of these enzyme classes later in this Section.

It is possible under *extreme* conditions of exercise, that the supply of electrons to CCO is too slow to prevent the release of significant quantities of partially reduced oxygen compounds. When this happens, cell damage can occur. Such a release of toxic molecules is known as a **respiratory burst**.

One more point about CCO is of interest. Cyanide is very effective at binding to haem groups, and it is likely that cyanide binds irreversibly to the haem group in CCO. Since the amount of CCO is very small compared to other haemproteins like myoglobin and haemoglobin, CCO is very susceptible to cyanide poisoning. In fact, it is thought that the toxicity of cyanide is due to its ability to inhibit CCO function. Blockage of CCO by cyanide stops respiration immediately and the cell dies, leading quickly to organism death.

6.2 Cytochrome P450

So far we have looked at proteins that bind O_2 for use in respiration; this is the major use of O_2 in the body. But O_2 is also used as an oxidising agent in a variety of oxidation reactions carried out by enzymes. One particular class of these enzymes is known as the cytochrome P450s[G] (when carbon monoxide is added to these particular enzymes, a carbon monoxide complex is formed, which has a strong absorption in the visible region at 450 nm; hence the name P450) and they catalyse the following reaction (where R is an alkyl group in a biomolecule):

$$\text{R–H(aq)} + O_2(g) + 2H^+(aq) + 2e^- = \text{R–OH(aq)} + H_2O(l) \qquad \mathbf{17}$$

What is remarkable about this reaction is that alkyl groups are rather inert. Certainly, reaction of common alkanes (e.g. propane, hexane) with air at room temperature is very slow, if not imperceptible. The cytochrome P450s (abbreviated to P450s hereafter) are powerful catalysts, capable of catalysing reaction 17, and related oxidation reactions, at body temperature in aqueous solution and at relatively low partial pressures of O_2. Moreover, this reaction is carried out by the enzyme stereospecifically; in other words only one enantiomer is produced when the product is chiral. Some examples of the reactions catalysed by P450s are shown in reactions 18–21:

$$\text{C}_6\text{H}_6 + O_2 + 2H^+ + 2e^- \longrightarrow \text{C}_6\text{H}_5\text{OH} + H_2O \qquad 18$$

$$(\text{CH}_3)_2\text{C}=\text{C}(\text{CH}_3)_2 + O_2 + 2H^+ + 2e^- \longrightarrow \text{epoxide} + H_2O \qquad 19$$

$$\text{R}_3\text{N} + O_2 + 2H^+ + 2e^- \longrightarrow \text{R}_3\text{N}^+\text{—O}^- + H_2O \qquad 20$$

$$\text{(CH}_3\text{)}_2\text{S} + O_2 + 2H^+ + 2e^- \longrightarrow \text{(CH}_3\text{)}_2\text{S}=O + H_2O \qquad 21$$

As reactions 18–21 show, the O_2 molecule is split in the reaction, with one oxygen atom ending up in a water molecule and the other in the product. Since only one of the oxygen atoms ends up in the product, P450s are known as '***mono*-oxygenases**'. (Other enzymes also use O_2 as a substrate[G], and in these cases both oxygen atoms end up in the product. Such enzymes are called '*di*-oxygenases'[G]; we shall not cover dioxygenases in any detail in this Block.)

Several individual types of P450 exist, but all are thought to have the same basic higher-order protein structure. Camphor 5-mono-oxygenase (a P450) has been crystallised and its crystal structure solved. The structure of part of the active site is shown in Figure 58.

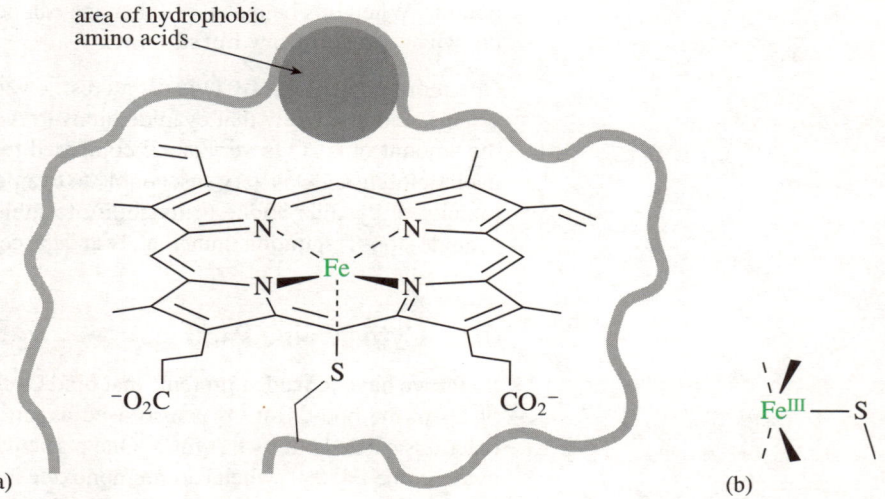

Figure 58 (a) Sketch of cytochrome P450 active site, showing hydrophobic pocket where the organic substrate binds; (b) condensed structure of the iron coordination geometry at the active site.

☐ How would you describe the structure of the active site?

■ The active site contains a haem group, the iron atom of which is also coordinated by the sulphur atom of a cysteinyl side-chain.

☐ Which group have we seen coordinating in this position in other proteins?

■ In other proteins we have only seen a histidyl side-chain in this position.

On the opposite side of the haem to the sulphur atom is a cavity, which is formed by a rigid arrangement of hydrophobic amino acids (such as valine, leucine). This is an ideal binding site for hydrophobic substrates, like fatty acids.

The proposed catalytic cycle for P450s is shown in Figure 59 (although there is still considerable debate about whether this cycle is correct). In its resting state (structure **22**, Figure 59) the iron is in its +3 oxidation state and the enzyme is inactive. The first step involves electron transfer from the outside of the protein to the active site, which reduces iron(III) to iron(II). (The process of electron transfer to the active site is very complicated and beyond the scope of this Course. The electron is obtained from a small molecule that acts as an electron carrier. The electron is then transferred from the surface of the protein through an electron transfer pathway, made up of amino acids, to the active site.) In the iron(II) state (structure **23**, Figure 59) the iron atom can bind an O_2 molecule (structure **24**, Figure 59). This is quickly followed by addition of a proton and a further electron to give a hydroperoxide ligand coordinated to iron(III) (structure **25**, Figure 59). A further proton then reacts with the hydroperoxide to give a water molecule and a highly reactive porphyrin–iron(V)–oxo group (structure **26**, Figure 59).

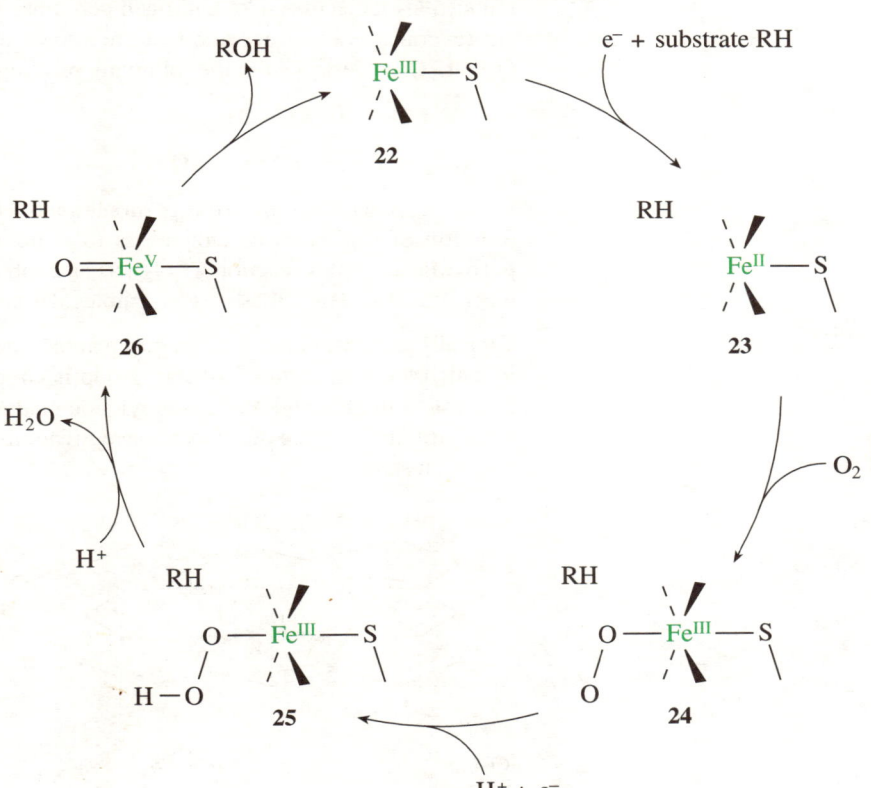

Figure 59 Catalytic cycle for the oxidation of organic molecules by cytochrome P450: structure **22**, resting state of enzyme; structure **23**, reduced state of enzyme before O_2 binding; structure **24**, after oxygen binding; structure **25**, hydroperoxide form; structure **26**, formation of reactive iron(V)–oxo species.

☐ How does this differ from the sequence of reactions for CCO?

■ We have seen a similar sequence for CCO (Figure 57), but in that case a porphyrin–iron(IV)–oxo species was generated, which rapidly reacted with two protons and further electrons to give a water molecule and iron(II).

In contrast to CCO, it is proposed that in P450s the highly reactive porphyrin–iron(V)–oxo reacts with an organic molecule (denoted as RH in the Figure), which is already bound in the active site adjacent to the haem group. Here we see the role of the hydrophobic cavity: the cavity binds the organic substrate, and, in so doing, brings the organic substrate in close proximity to the iron(V)–oxo group. The consequence is that when the iron(V)–oxo group is generated, it can react quickly with the organic substrate before it can be reduced to water with further electrons and protons. The last step of the cycle produces the monooxygenated substrate, ROH, and regenerates the enzyme in its resting state (structure **22**, Figure 59).

☐ What other proteins will bind O_2 when they are in the iron(II) oxidation state?

■ Haemoglobin, myoglobin and haemerythrin all contain iron(II) in their deoxygenated form. On oxygenation the iron(II) is formally oxidised to iron(III).

Note that P450s cannot bind O_2 in the iron(III) state and require an electron before becoming active. In the same way, if myoglobin or haemoglobin are oxidised to iron(III) (not by O_2) to give a deoxygenated iron(III) form, then this form cannot bind oxygen. Cooking red meat is an example of this! Fresh, uncooked meat is bright red. The colour is due to oxygenated haemoglobin and myoglobin. On standing (or cooking) the meat turns brown. The brown colour is the inactive iron(III) form of the proteins, with a water molecule coordinated to the iron of the haem group instead of O_2.

Turning back to the catalytic cycle of CCO, you will see parallels between the mechanisms of CCO and P450. The most important similarity is the proposed generation of a highly reactive iron–oxo species, either iron(IV)–oxo or iron(V)–oxo. It is the high reactivity of this species that allows both enzymes to react very quickly and efficiently.

6.3 Peroxidases, catalases and Cu–Zn superoxide dismutase

As we saw with CCO, respiration reduces O_2 usually to water, but sometimes to partially reduced O_2 molecules, like hydrogen peroxide. To address this potential problem, there are several classes of enzyme that are known to catalyse the destruction of H_2O_2 and $O_2^{\cdot -}$. H_2O_2 is destroyed in the following reactions (where X is an organic molecule):

$$2H_2O_2 = 2H_2O + O_2 \qquad 18$$

$$H_2O_2 + XH_2 = X + 2H_2O \qquad 19$$

Reaction 18, which is the disproportionation of H_2O_2, is catalysed by **catalases**. The sole function of this class of enzyme is to remove H_2O_2. Reaction 19 is catalysed by **peroxidases**; in this reaction, H_2O_2 acts as a substrate to oxidise an organic molecule (in biochemical systems, these organic molecules can be fatty acids, phenols, amines, etc.).

Typically, catalases have a haem group at their active sites. Figure 60 shows an example in which the iron atom of a haem group is coordinated by a tyrosinate* side-chain (in contrast to the histidyl and cysteinyl side-chains that we have seen in this position in other proteins). Unlike in deoxygenated myoglobin, for example, here the iron is in its +3 oxidation state.

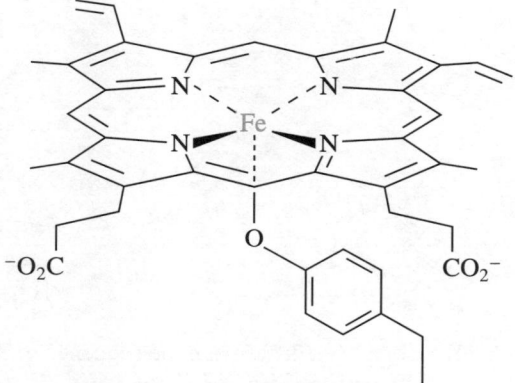

Figure 60 The active site of a catalase.

The mechanism of catalase-mediated hydrogen peroxide disproportionation involves two steps:

$$H_2O_2 + \text{enzyme–iron(III)} = H_2O + \text{enzyme intermediate} \qquad 20$$

$$\text{enzyme intermediate} + H_2O_2 = H_2O + O_2 + \text{enzyme–iron(III)} \qquad 21$$

$$\text{Overall reaction: } 2H_2O_2 = 2H_2O + O_2 \qquad 18$$

SAQ 17 Sketch the structure of the active site of the enzyme intermediate in steps 20 and 21.

SAQ 18 Knowing that peroxidases also have a porphyrin–Fe(III) group at their active sites, propose a mechanism of action for peroxidases. (Bear in mind the catalase mechanism discussed above.)

* Tyrosine anion.

We have already encountered an enzyme called copper–zinc superoxide dismutase (abbreviated to Cu–Zn SOD hereafter) in Section 3. This enzyme catalyses the following reaction:

$$2O_2^- + 2H^+ = O_2 + H_2O_2 \qquad 22$$

In Section 5 (Figure 24) we saw that the O_2^- ion (superoxide ion) is a radical molecule (see also Figure 26). It is extremely reactive and will indiscriminately react with cell components. Even trace quantities of superoxide, which might come from incomplete reduction of O_2 by CCO, are hazardous. Cu–Zn SOD is a very efficient enzyme, which catalyses the reaction of superoxide to form the less harmful O_2 and H_2O_2.

☐ How will a system then deal with hydrogen peroxide?

■ H_2O_2 is later removed by catalases or peroxidases.

The active site does not contain a haem group and the Cu–Zn SOD's mechanism of action does not involve an iron–oxo species. A proposed mechanism is shown in Figure 61. The Figure shows that the copper is the active metal, changing between oxidation states I and II during the catalytic cycle. In the first few steps, superoxide binds to the copper, which is then reduced to copper(I) with the concomitant release of O_2. The bridging imidazolate is also protonated in these steps (structures **27–29**, Figure 61). In the next step, another superoxide molecule and a proton enter the active site. The superoxide coordinates to the copper(I) and also hydrogen bonds to two protons (structure **30**, Figure 61). The copper(I) reduces the superoxide to peroxide, which is then released as hydrogen peroxide, and is itself reoxidised to copper(II).

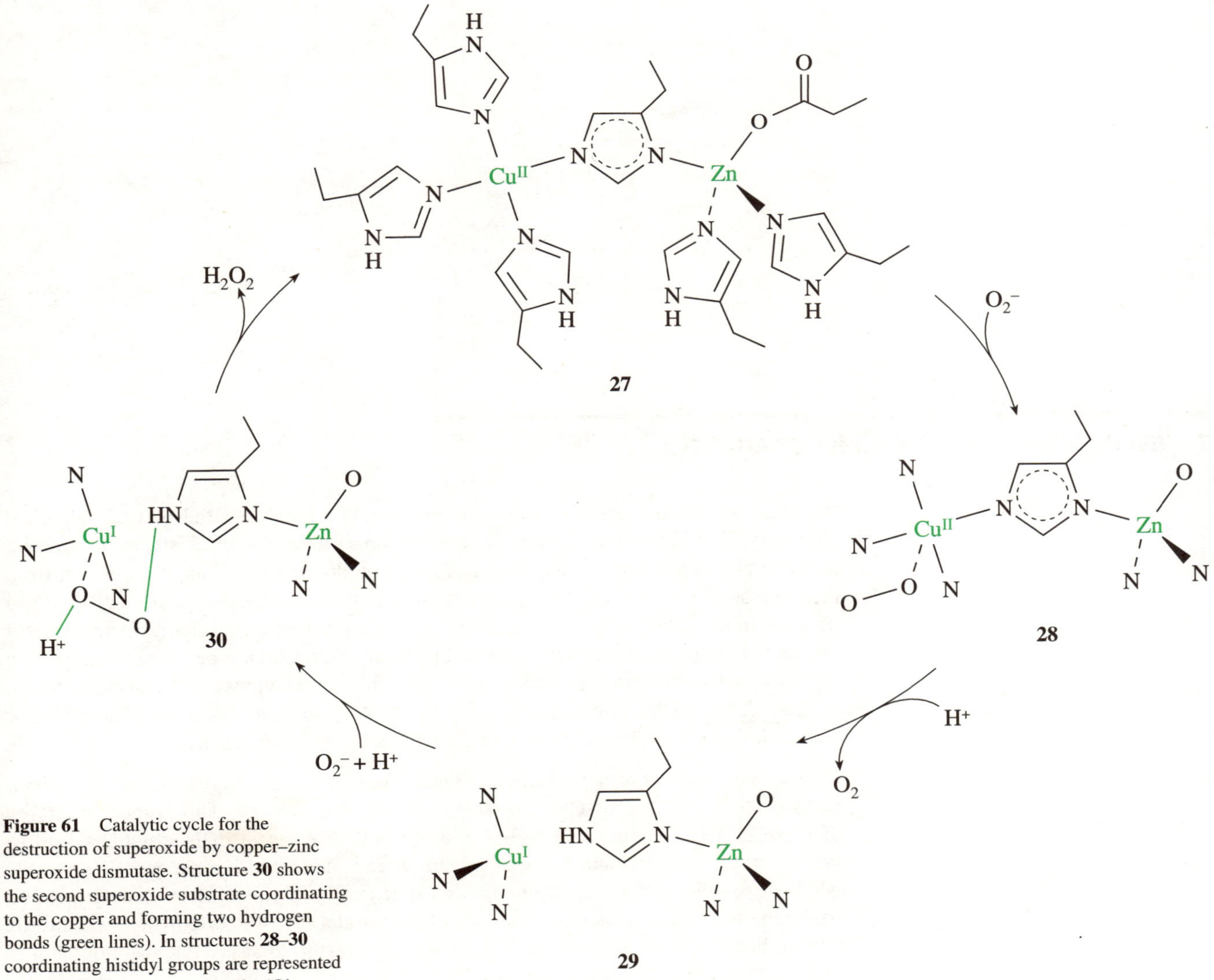

Figure 61 Catalytic cycle for the destruction of superoxide by copper–zinc superoxide dismutase. Structure **30** shows the second superoxide substrate coordinating to the copper and forming two hydrogen bonds (green lines). In structures **28–30** coordinating histidyl groups are represented by 'N' and the aspartate group by 'O'.

6.4 Summary of Section 6

In this Section we have examined the structures and mechanisms of several enzymes that activate O_2-containing molecules. All the enzymes are involved in the control of O_2 within the body, with the main aim of reducing the concentration of harmful, partially reduced O_2 molecules. A common feature of the haem enzymes is the occurrence of a reactive porphyrin–iron–oxo species in the catalytic cycle. The high reactivity of these species ensures that the reactions at the active site are rapid and, in most cases, complete. The key points are:

1 Cytochrome c oxidase catalyses the reduction of O_2 to H_2O in the last step of aerobic respiration.

2 Cytochrome P450s catalyse the mono-oxygenation of alkyl and other functional groups.

3 Catalases and peroxidases are important in the destruction of H_2O_2.

4 Copper–zinc superoxide dismutase is important in the destruction of the superoxide radical.

5 Porphyrin–iron–oxo species are common intermediates in the catalytic cycles of haem enzymes.

SAQ 19 Catalases and superoxide dismutases are among the most efficient enzymes known. Why is this important?

SAQ 20 Describe in as much detail as possible the mechanism of the oxidation of camphor (**31**) to 5-hydroxycamphor (**32**) by camphor-5-mono-oxygenase.

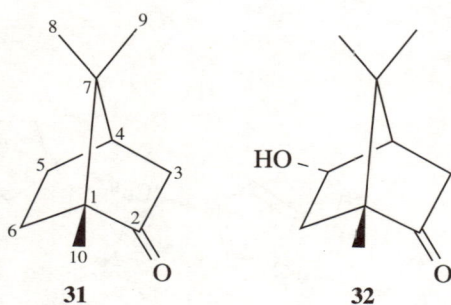

7 IRON TRANSPORT AND STORAGE

We have seen in the previous Sections that metals are an essential part of biological chemistry. Of all the trace elements, iron is the most important, especially as it is present in many essential enzymes and proteins. But how do organisms acquire the iron from their surroundings? Clearly, organisms need to absorb iron biochemically before it can be used in proteins. Also, some method of replacing lost iron quickly is needed; for instance, how is blood replaced once it has been lost through a cut? This prompts the question: what biochemical systems are responsible for the uptake and transport of iron, or, indeed, any metal, within an organism? In this Section we shall examine something of what is known about iron uptake, transport and storage in organisms.

We begin by looking back to Section 5, where we saw that the iron proteins, myoglobin and haemoglobin, are essential for O_2 transport and storage. The higher structures of these proteins are very precise, and any small changes in their structures lead to inefficient O_2 transport and storage. Indeed, there is a class of human genetic diseases, which affects the structure of haemoglobin specifically. In this class of diseases, called **thalassaemia**[G], the molecular structure of haemoglobin is distorted (the structure of normal haemoglobin is shown in Figure 40); as a result, thalassaemic haemoglobin is inefficient at transporting O_2 (see Box 2).

Box 2 Thalassaemia

Thalassaemia is one of the most common of hereditary (i.e. genetic) human diseases. Thalassaemia patients suffer from similar symptoms to anaemic patients, as essentially both have low counts of healthy red blood cells. The disease is called thalassaemia (Greek: thalassa, sea) because it is very common in countries surrounding the Mediterranean Sea. It is also widespread in Central Africa, India and South-East Asia (Figure 62). Thalassaemia is also known as Cooley's anaemia, after the American physician who first identified it. There are two main types of the disease, thalassaemia-minor and thalassaemia-major. The former usually has symptoms of mild anaemia, whereas the latter is seriously debilitating.

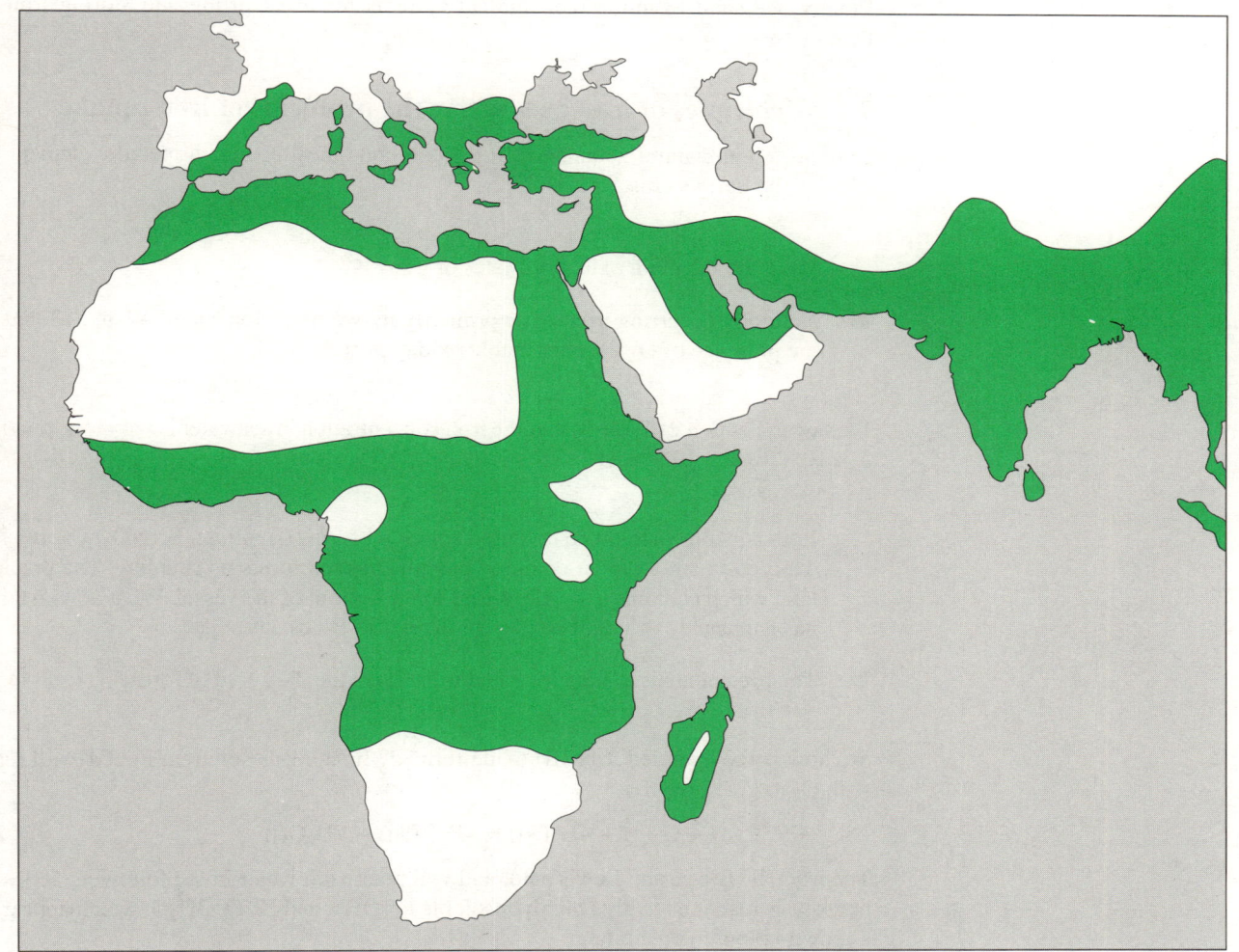

Figure 62 Areas of high incidence of thalassaemia.

The treatment for thalassaemia-major is regular blood transfusions, which restore the healthy red-blood cell count to normal levels. However, this leads to problems. The problems are not directly linked to the symptoms of anaemia, which are largely ameliorated by the transfusions, but to the regular intake of large quantities of iron into the body. The human body is capable of excreting a maximum of about 10 mg of iron a day, whereas regular blood transfusions put far more iron into the body. This leads to a condition known as **iron overload** or **haemochromatosis**[G].

Iron, like all other metals in an organism, has an optimal concentration: too little or too much leads to ill health, as shown in Figure 2. For example, one of the symptoms of iron overload is increased susceptibility to bacterial infection. This extra iron can be absorbed by other organisms, particularly bacteria (remember that not only humans require iron to live), which then proliferate and cause infection.

Due to its high mortality rate and widespread occurrence, thalassaemia is the subject of much medical research. It does, however, highlight some important principles for the bioinorganic chemist. Firstly, it is clear that within the healthy human body, biochemical systems carefully control the level of iron so that it is around an optimal value (this control is known as **homeostasis**). Secondly, it appears as if the human body has a means of transporting iron and probably storing it. For example, some iron stores are required in case of sudden loss of iron, say from heavy bleeding. Thirdly, within the body, iron is not made available to other organisms, thus helping to reduce the possibility of bacterial infection. Fourthly, bacteria themselves must have a means of absorbing iron.

In this Section we shall examine the principles stated above. To begin with, it will be necessary to study some of the basic chemistry of iron and its coordination complexes. Then we shall look at the biochemical systems involved in iron uptake by bacteria. Finally, we shall examine biochemical systems for transporting and storing iron in humans.

7.1 Principles of iron chemistry: the problems of iron uptake

Iron has a high natural abundance. It is the second most abundant metallic element by mass in the Earth's crust (7.1%).

☐ What are the main oxidation states of iron?

■ Naturally occurring iron exists primarily in two oxidation states, +2 and +3, but in the presence of O_2 the most stable oxidation state is +3.

We saw in Section 2 (Table 3) that the iron concentration in seawater is very low, roughly 5×10^{-11} to 2×10^{-8} mol l^{-1}.

☐ If an aqueous pale-green coloured solution of iron(II) nitrate, $Fe(NO_3)_2$, at pH 7 is exposed to air, a brown-coloured precipitate soon forms on standing. The precipitate, which contains iron(III), settles to the bottom of the vessel. What do you think has happened? (It might be helpful to refer to Block 1.)

■ The precipitate is mostly hydrated iron(III) oxide, $Fe_2O_3.nH_2O$ (rust).

So, how is the hydrated iron(III) oxide formed? In aqueous solution, iron(II) will react with O_2 to give iron(III):

$$4Fe^{2+}(aq) + O_2(g) + 4H^+(aq) = 4Fe^{3+}(aq) + 2H_2O(l) \qquad 23$$

The iron(III) is a strong Lewis acid, and will react with water in the following series of hydrolysis reactions to give highly insoluble $Fe(OH)_3$ and $Fe_2O_3.3H_2O$, which appear as a brown-coloured precipitate:

$$Fe^{3+}(aq) + H_2O(l) = [Fe(OH)]^{2+}(aq) + H^+(aq) \qquad 24$$

$$[Fe(OH)]^{2+}(aq) + H_2O(l) = [Fe(OH)_2]^+(aq) + H^+(aq) \qquad 25$$

$$[Fe(OH)_2]^+(aq) + H_2O(l) = Fe(OH)_3(s) + H^+(aq) \qquad 26$$

$$2Fe(OH)_3(s) = Fe_2O_3.3H_2O(s) \qquad 27$$

Notice that the hydrolysis reactions *must* be pH dependent, since a proton is produced in each of the reactions. Therefore, at low pH, say less than pH 1.5, iron(III) is soluble in water, but at pH 7 insoluble $Fe(OH)_3$ is formed. $Fe(OH)_3$ is profoundly insoluble, with a solubility product, $K_{sp} = 2 \times 10^{-39}$ mol^4 l^{-4}.

☐ What is the concentration of Fe^{3+}(aq) in H_2O at pH 7, assuming that $Fe(OH)_3$(s) is present?

- We know that the solubility product, K_{sp}, of $Fe(OH)_3 = 2 \times 10^{-39} \text{ mol}^4 \text{ l}^{-4}$. Therefore,

    $[Fe^{3+}][OH^-]^3 = 2 \times 10^{-39} \text{ mol}^4 \text{ l}^{-4}$

    At pH 7: $[H^+] = 10^{-7} \text{ mol l}^{-1}$ and $[OH^-] = 10^{-7} \text{ mol l}^{-1}$. (Remember that we calculate this from $K_w = [H^+][OH^-] = 10^{-14} \text{ mol}^2 \text{ l}^{-2}$.) Therefore,

    $$[Fe^{3+}] = \frac{2 \times 10^{-39}}{(10^{-7})^3}$$

    $$= 2 \times 10^{-18} \text{ mol l}^{-1}$$

From the above calculation, we can see that at pH 7, the concentration of $Fe^{3+}(aq)$ is extremely small: $2 \times 10^{-18} \text{ mol l}^{-1}$. The value of 5×10^{-11} to $2 \times 10^{-8} \text{ mol l}^{-1}$ for $Fe^{3+}(aq)$ in seawater, given in Table 3, is much higher than this.

- ☐ Why do you think the concentration of $Fe^{3+}(aq)$ in seawater is higher than the value you have just calculated?

- ■ This is due to the presence of other ligands that can form soluble iron complexes. For example, chloride, bromide, acetate and nitrate are all potential ligands.

Nevertheless, for such an important bioinorganic element, the concentration of $Fe^{3+}(aq)$ in neutral aqueous solution is very low indeed. This very low concentration is a significant problem when it comes to an organism obtaining iron, since iron in its soluble form is in such short supply. If an organism is to survive, it must have some biochemical means of absorbing the trace amounts of surrounding iron. Also, the organism must prevent iron(III) oxide precipitation once the iron is inside the organism, since this may lead to cell damage. (Although having said this, some organisms, such as pigeons, are known to crystallise iron oxide deliberately within certain parts of their bodies, particularly the brain. These small iron oxide pellets are often magnetic and are thought to act as in-built compasses.)

- ☐ What are the products of the reaction of a single iron(II) ion with a single molecule of O_2?

- ■ $Fe^{II} + O_2 = Fe^{III} + O_2^{-\cdot}$

    The reaction of O_2 with iron(II) gives single-electron oxidation of the iron(II) to iron(III) and the generation of a superoxide radical anion.

From Section 5 we know that this anion is highly toxic to biological systems and must be avoided where possible. Therefore, any free iron(II) is potentially detrimental, acting as a reducing agent for O_2.

Free iron(II) dissolved within an organism is also potentially dangerous. Therefore, the organism must have methods for preventing the formation of free iron(II). (*Note* It is now easy to see why iron-overload patients suffer from a variety of secondary diseases, since the extra iron is not controlled by the normal biochemical systems that deal with iron. There is an increased availability of free iron, which can be chelated by micro-organisms, and an elevated level of iron(II).)

The last property of iron we shall examine is the thermodynamic stability of its coordination complexes.

- ☐ Would you classify iron(III) as a hard or soft metal?

- ■ Iron(III) is a first-row transition metal in a high oxidation state, and so is classified as a hard metal. As such, it will tend to form stable complexes with hard ligands (Block 4). Typical hard ligands contain oxygen and nitrogen as the coordinating atoms.

☐ Would you expect iron(II) or iron(III) to make a more stable complex with the edta^{4-} anion (Figure 63a)?

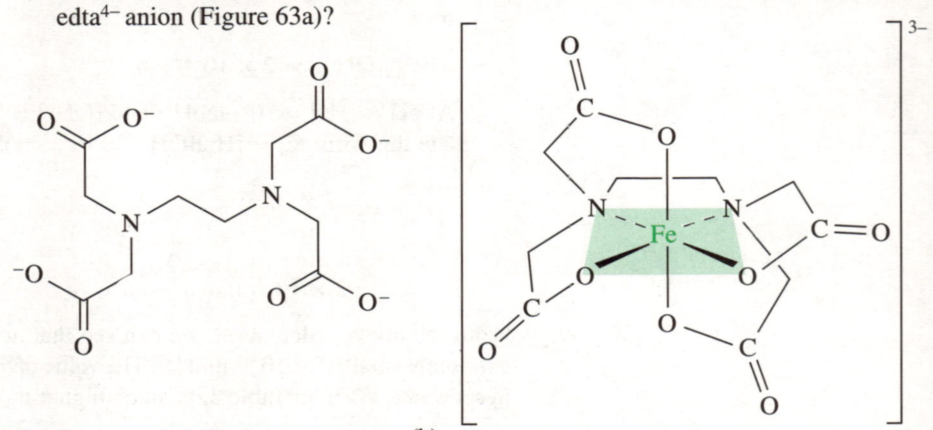

Figure 63 (a) The edta^{4-} anion; (b) its coordination mode to iron.

■ The hexadentate edta^{4-} ligand coordinates through both O and N. The stability constant^G of the [iron(III)–edta]$^{3-}$ complex (shown in Figure 63b) is c. 10^{25} mol^{-1} l, whereas the analogous stability constant with iron(II) is only c. 10^{14} mol^{-1} l.

Complexes containing chelate rings are usually more thermodynamically stable than similar complexes without rings; for instance, $\log_{10}$ (stability constant) for [Ni(NH$_3$)$_6$]$^{2+}$ is 8.61, whereas that for [Ni(en)$_3$]$^{2+}$ is 18.18. This is known as the **chelate effect**. It is observed for pairs of complexes when the coordinating atom of a monodentate ligand, L, is the same as that of a bidentate ligand, L—L, and there is no steric strain in the chelate ring. It is thought that several factors are involved overall in the chelate effect, but that the most influential is the entropy change in the formation reaction. This can be seen in the experimental data given below in Table 8 for the formation of two four-coordinate cadmium complexes:

$$\text{Cd(aq)}^{2+} + n\text{L} = [\text{Cd(L)}_n]^{2+} \qquad \textbf{28}$$

[Cd(CH$_3$NH$_2$)$_4$]$^{2+}$, **33**, is coordinated through nitrogen to four monodentate methylamine ligands, whereas [Cd(en)$_2$]$^{2+}$, **34**, is coordinated through nitrogen to two bidentate ethylenediamine ligands. We see that the chelated complex is far more stable than the complex with no rings — by a factor of over 10^4 in this case.

Table 8 Stability constants and thermodynamic data at 298.15 K for some cadmium(II) complexes

Complex	Log$_{10}$ (stability constant, K)	$\Delta H_m^\ominus$ /kJ mol^{-1}	$\Delta G_m^\ominus$ /kJ mol^{-1}	$\Delta S_m^\ominus$ /J K^{-1} mol^{-1}	$-T\Delta S_m^\ominus$ /kJ mol^{-1}
[Cd(CH$_3$NH$_2$)$_4$]$^{2+}$	6.55	−57.32	−37.41	−66.8	19.91
[Cd(en)$_2$]$^{2+}$	10.62	−56.48	−60.67	14.1	−4.19

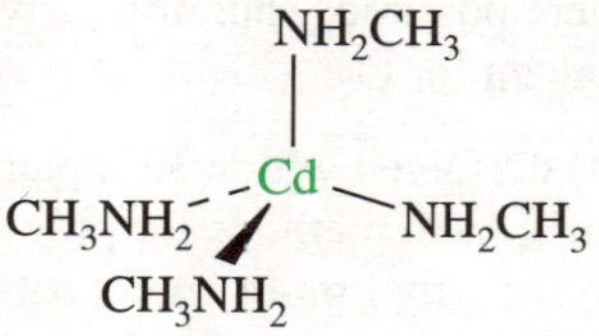

33

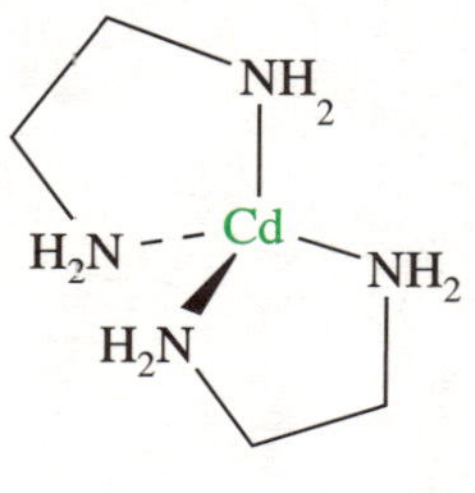

34

It is evident that the enthalpy change for the formation reaction of each complex is very similar, which is to be expected because the atoms involved in each bond, and therefore type of bonding, are similar in each case.

☐ What do you notice about the values for the change in entropy?

■ They are very different. The entropy change for the chelated complex is positive, whereas that for the complex with no chelate rings, is negative.

The large difference in entropy changes, $\Delta S_m^\ominus$, leads to a big difference in the Gibbs free energy changes for the reactions.

☐   How does this lead to a difference in the stability constants of the two complexes?

■   From the relationship $\Delta G_m^\ominus = \Delta H_m^\ominus - T\Delta S_m^\ominus$, we see that a *positive* entropy change will lead to a lower or more negative value for the Gibbs free energy change $\Delta G_m^\ominus$, of the reaction. $\Delta G_m^\ominus = -2.303RT \log_{10} K$, so the more negative is $\Delta G_m^\ominus$, then the larger is K and the more stable is the complex.

The leading question now is, why does this happen? The answer seems to be that the dominating factor is the entropy of the ligands. Consider what happens when a hexa-aquo metal ion, M, reacts with a complexing monodentate ligand, L:

$$[M(H_2O)_6]^{n+} + 6L = [ML_6]^{n+} + 6H_2O \qquad 29$$

there is no change in the number of molecules in solution: the six water molecules are displaced by six ligand molecules.

☐   What is the significant difference when the same reaction takes place with bidentate ligands?

■   In a similar reaction with bidentate ligands, L–L,

$$[M(H_2O)_6]^{n+} + 3L\text{–}L = [M(L\text{–}L)_3]^{n+} + 6H_2O \qquad 30$$

there is a net increase of three molecules in solution.

SLC 6   Now from a Second Level Course we know that for gas–solid reactions, a greater increase in the number of gas molecules in a reaction leads to a greater entropy change. Something similar happens here, but in solution. The increase in the number of independent molecules in the chelate reaction leads to a more positive $\Delta S_m^\ominus$.

The chelate effect is usually at a maximum for five- and six-membered rings: smaller rings tend to suffer from strain, and in larger rings the second coordinating atom is no longer very close to the metal.

Macrocyclic ligands, such as porphyrin rings, tend to show the same stabilising effect as polydentate ligands and this is sometimes known as the **macrocyclic effect**.

7.1.1   Summary of iron chemistry

Try to summarise the main points of iron chemistry for yourself, and then compare it with the list below.

1   From a bioinorganic point of view, we can conclude that despite the high natural abundance of iron, it is in scarce supply in water.

2   The more stable form of iron in oxygenated water is iron(III).

3   The low concentration of Fe^{3+} in water is because of the extreme insolubility of $Fe(OH)_3$.

4   Any iron absorbed by an organism is potentially detrimental in its +2 oxidation state, due to its reaction with O_2 to give superoxide radical anions.

5   Iron(III) is coordinated preferentially by hard ligands, especially hard polydentate ligands, which give iron(III) complexes with very high stability constants.

These chemical properties of iron demand that biochemical systems have efficient and effective methods of obtaining and controlling iron. Specifically, they must be able to (i) solubilise and assimilate iron in their local environment, and (ii) protect the iron once it has been absorbed. The rest of this Section will describe how this has been achieved by some organisms.

Before moving on, it is worth noting that there is a constant competition for the available iron in the natural world. In many cases the availability of iron is the determining factor in whether an organism can proliferate or not. For example, it is believed that the low concentration of iron in seawater limits the amount of plankton growth.

7.2 Iron uptake by organisms

Nearly all organisms are able to take up iron. However, only a handful of organisms have had their iron-uptake chemistry studied. The organism that has received most attention (other than human) is a single-cell, prokaryotic^G bacterium (found in the human large intestine and elsewhere), called *Escherichia coli*^G (abbreviated to *E. coli*), a high-resolution image of which is shown in Figure 64. The reason that this bacterium has been so thoroughly studied is that it is relatively easy to grow and study colonies of it in the laboratory. The iron-uptake mechanism in *E. coli* is known in a fair degree of detail.

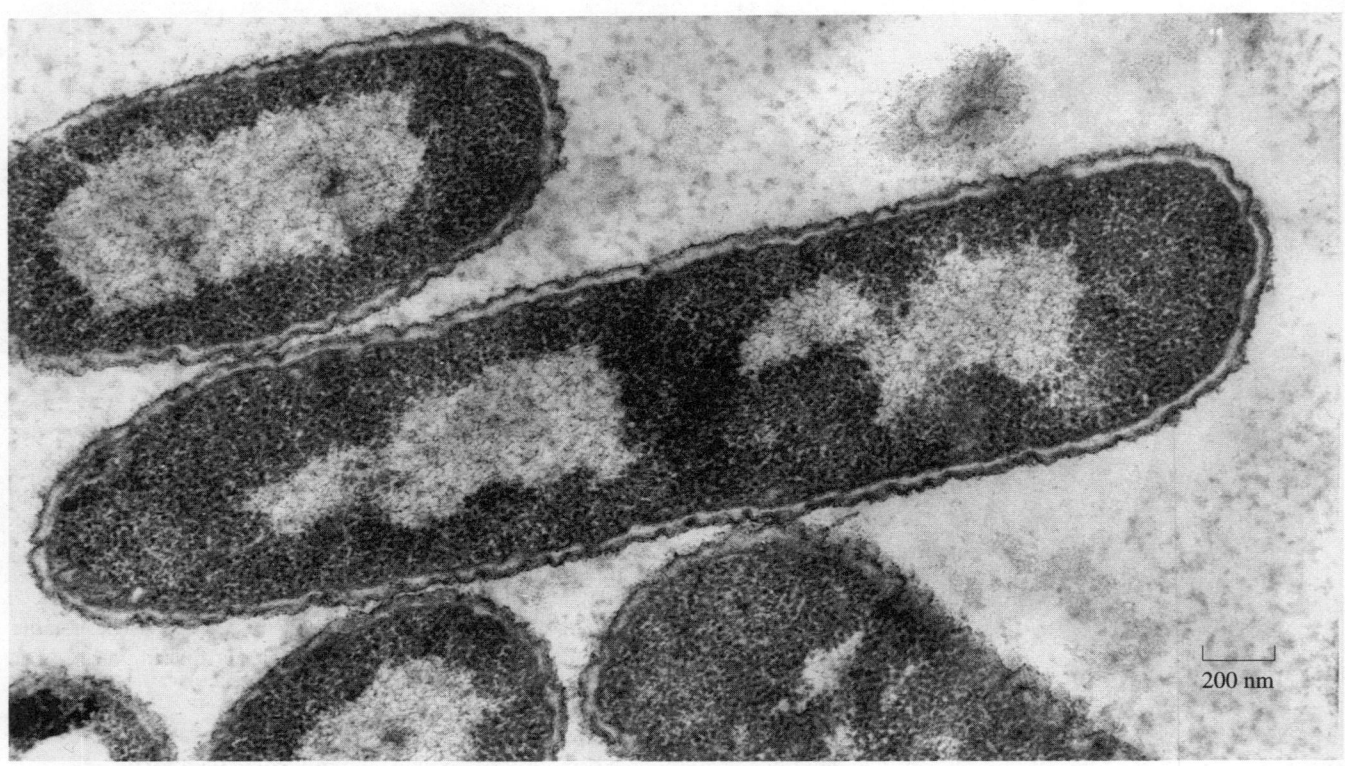

Figure 64 There are many harmless strains of the *E. coli* bacterium; the ones found naturally in the human gut are useful because they synthesise several vitamins of the B-complex and vitamin K. However, there are also over 100 pathogenic (i.e. disease-causing) strains of the bacteria. The most infamous is probably *E. coli* O157:H7, which is very virulent. This strain can find its way into the human food chain (from the intestines of cattle where it is thought to originate), and it causes severe food poisoning due to the toxins excreted by the bacteria. The toxins are absorbed from the gut into the bloodstream; damage to the kidneys occurs, which may eventually result in death, particularly for very old or young persons.

E. coli obtains its iron in a remarkable fashion. Each *E. coli* bacterium within a colony, secretes small molecules that are capable of specifically chelating iron. These small molecules are known as **siderophores**^G (from the Greek for iron carriers; pronounced 'sid-air-o-fores'; see also Block 3). Several types of siderophore are known, each capable of chelating iron in a stable iron–siderophore complex. The structures of some known siderophores are shown in Figure 65.

We shall examine the properties of one of the siderophores in more detail below. For all the siderophores, however, their *modus operandi* is to be secreted from the bacterium, to chelate an iron(III) ion selectively in a stable complex, and then to be re-absorbed by a bacterium (not necessarily the original bacterium) as the iron(III)–siderophore complex (see Figure 66 for a schematic representation).

☐ Derive an expression for the concentration of an iron(III)–siderophore complex in terms of its stability constant. Explain why an iron(III) siderophore complex needs to have a very high stability constant in order to be biochemically useful.

■ By writing the equation for the formation of the complex, we can then derive an expression for the concentration of an iron(III)–siderophore complex in terms of its stability constant:

$$\text{Fe}^{3+}(\text{aq}) + \text{siderophore}^{n-}(\text{aq}) = \text{Fe–siderophore}^{(3-n)+}(\text{aq}) \qquad 31$$

$$K_s = \frac{[\text{Fe–siderophore}^{(3-n)+}(\text{aq})]}{[\text{Fe}^{3+}(\text{aq})][\text{siderophore}^{n-}(\text{aq})]} \qquad 32$$

Rearranging gives:

$$[\text{Fe–siderophore}^{(3-n)+}(\text{aq})] = K_s[\text{Fe}^{3+}(\text{aq})][\text{siderophore}^{n-}(\text{aq})] \qquad 33$$

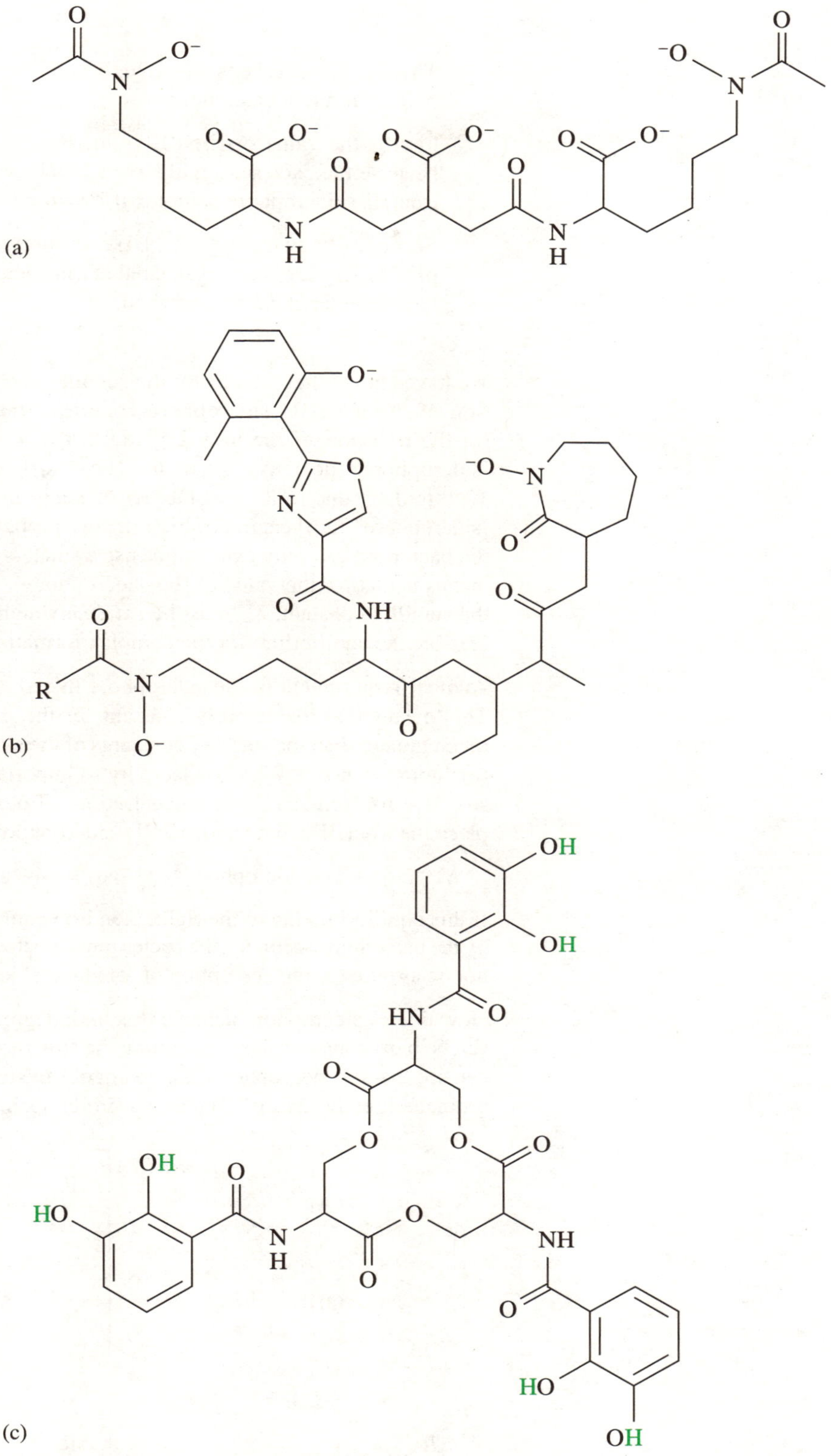

Figure 65 Structures of three siderophores: (a) aerobactin; (b) mycobactin; (c) enterobactin[G]; acidic hydrogens are printed in green.

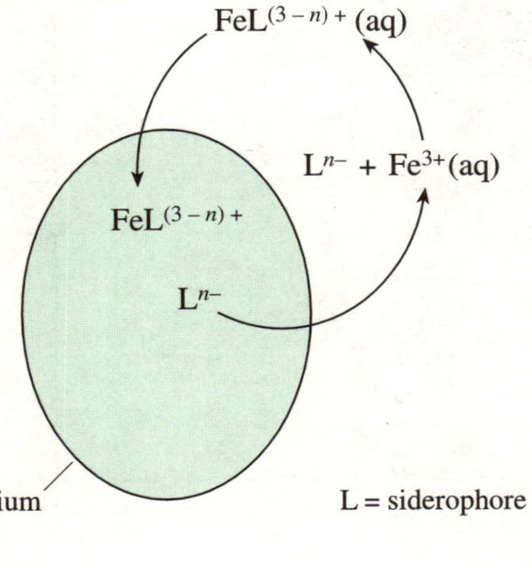

Figure 66 Schematic diagram of iron uptake by a siderophore.

There are two reasons why a high stability constant is required for efficient transport of iron to a bacterium:

Firstly, the value of [iron(III)–siderophore$^{(3-n)+}$(aq)] must be significant; this is because the bacterium will have a much better statistical chance of absorbing the iron(III)–siderophore complex if it is in relatively high concentration.

Secondly, knowing that iron(III) is in short supply and its concentration in water at pH 7 is very low, any organism that can competitively chelate the available iron will have a better chance of survival.

We have shown that the greater the stability constant, the higher will be the concentration of the iron(III)–siderophore complex, thus providing the optimum conditions for the transport of the iron. Let's take a look at the rough values of [Fe^{3+}(aq)] and [siderophore^{n-}(aq)]. We know that [Fe^{3+}(aq)] cannot be high and may be as low as 10^{-18} mol l^{-1}, due to the insolubility of many iron(III) compounds. Also, the value of [siderophore^{n-}(aq)] cannot be high (that is, probably much less than 10^{-12} mol l^{-1}), since the bacterium can only ever synthesise a small amount of the siderophore. What this all means is that for the value of [Fe–siderophore$^{(3-n)+}$(aq)] to be significant, the value of the stability constant, K_s, must be *extremely* high. In other words, for this system to be feasible the equilibrium for the complex formation must lie very heavily to the right.

Another requirement of the siderophore ligand is that it must be selective for iron(III). This means that the stability constant for the iron(III)–siderophore complex must be much greater than the stability constants of the siderophore complexes with other metals (including iron(II)). Why is selectivity so important? Firstly, if the siderophore were not selective for iron(III), high concentrations of other metal ions (M^{m+}) would easily displace the iron(III) from any iron(III)–siderophore complex, according to the equation

$$M^{m+}(aq) + Fe\text{–siderophore}^{(3-n)+}(aq) = M\text{–siderophore}^{(m-n)+}(aq) + Fe^{3+}(aq) \quad 34$$

If this equilibrium lay to the right, then iron could not be obtained in any great amounts by the bacterium. Secondly, the bacterium's biochemical systems for absorbing iron should not be a route for the absorption of toxic metal ions, such as mercury and cadmium.

How does a siderophore achieve this high degree of iron(III) selectivity? The question can be partly answered by examining the structures of the siderophores in Figure 65. We can see that the siderophores are analogous to simpler organic molecules which can co-ordinate directly to iron(III) ion in a similar fashion, as indicated in reactions 35–37:

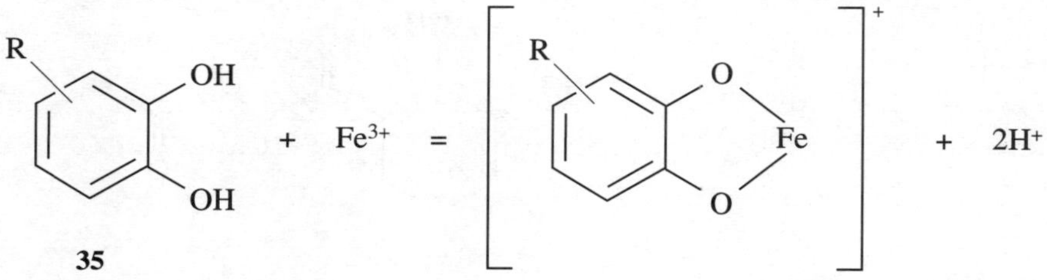

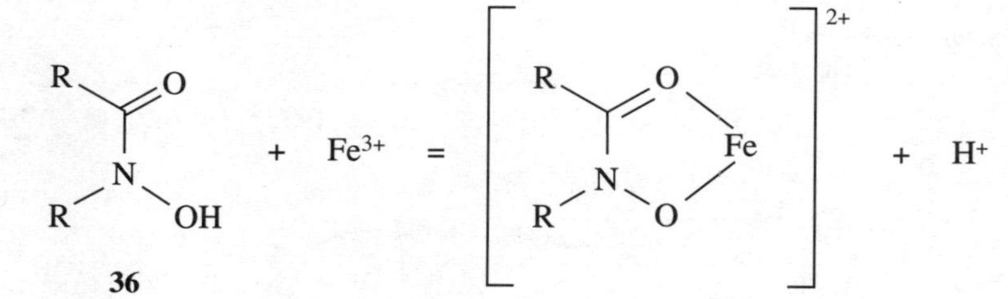

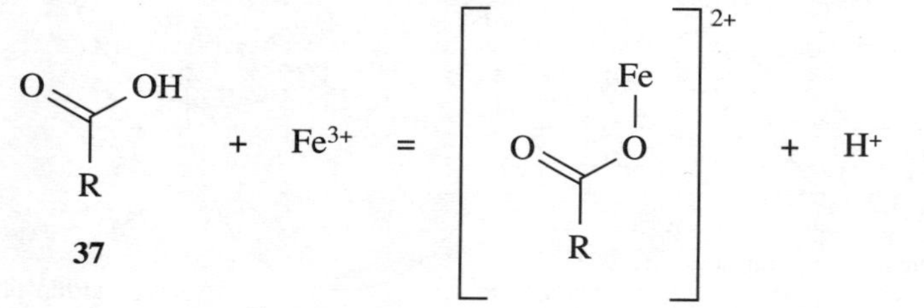

□ Are the coordinating groups of the ligands in reactions 35–37 hard or soft?

■ All the groups shown are hard ligands and, as such, form stable complexes with hard metals, like iron(III) and aluminium(III).

The groups in structures **35** and **36**, 1,2-dihydroxybenzene (trivial name catechol, pronounced 'kat-a-kol') and hydroxamic acids, respectively, form particularly stable complexes with iron(III), because they are chelating groups. The chelate 'bite' of these two groups is just about right to form a very stable complex with iron(III).

Of all the siderophores, the one that has received most attention is enterobactin, shown in Figure 65c. The reason for this attention is that enterobactin forms an exceedingly stable complex with iron(III); in fact, it is the most stable, soluble iron(III) complex that is known. The stability constant* of fully deprotonated enterobactin with iron(III) is extremely high at about 10^{49} mol^{-1} l!

What are the chemical and structural features of enterobactin that give it such a high stability constant with iron(III)? To answer this question we need to examine the structure of enterobactin in more detail. Figure 67 shows the structure of the iron(III)–enterobactin complex.

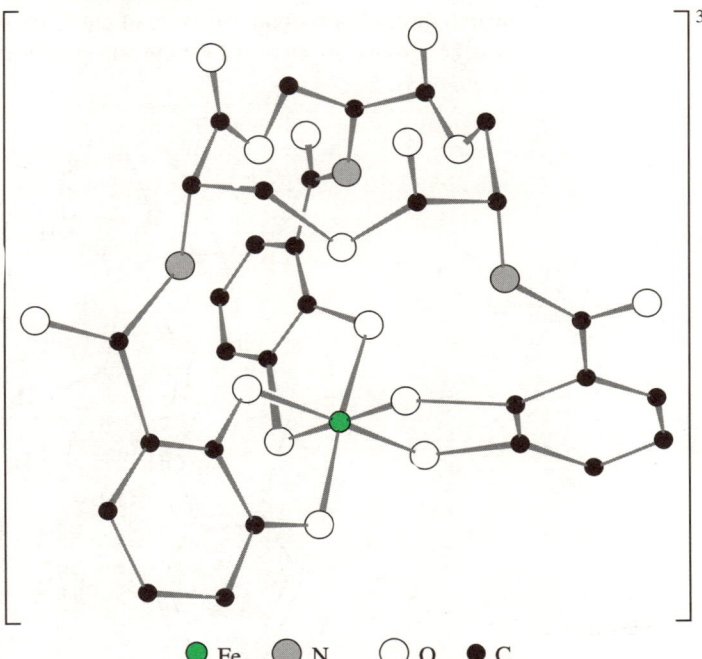

Figure 67 Three-dimensional structure of iron–enterobactin complex; the hydrogen atoms have been omitted for clarity.

□ Use your Orbit kit to make a model of the iron–enterobactin complex. Which type of group found in siderophore model compounds coordinates to the iron in the complex?

■ The iron atom is chelated by three, deprotonated catechol (known as catecholate) groups.

Notice that iron(III) complexation displaces the six protons on the catechol oxygen atoms of enterobactin, so, overall, enterobactin is a hexadentate ligand providing six coordinating atoms to the iron. The catechol rings are attached to each other via a twelve-membered ring of carbon and oxygen atoms. This ring is a serine trimer, condensed together as follows:

$$3\ \text{serine} \rightleftharpoons \text{triserine ring} + 3\text{H}_2\text{O} \qquad 38$$

* Note that the definition of stability constant (Block 3) assumes an equilibrium reaction in aqueous solution between a hydrated metal ion and ligand(s), such that we would write enterobactin in its ionised — that is, fully deprotonated — form.

Rather than forming the normal peptide C(O)—NH bond between the individual amino acid molecules, the serines are linked via ester bonds, whereby the ester is formed between the —OH of one serine side-chain and the —CO$_2$H group of another serine:

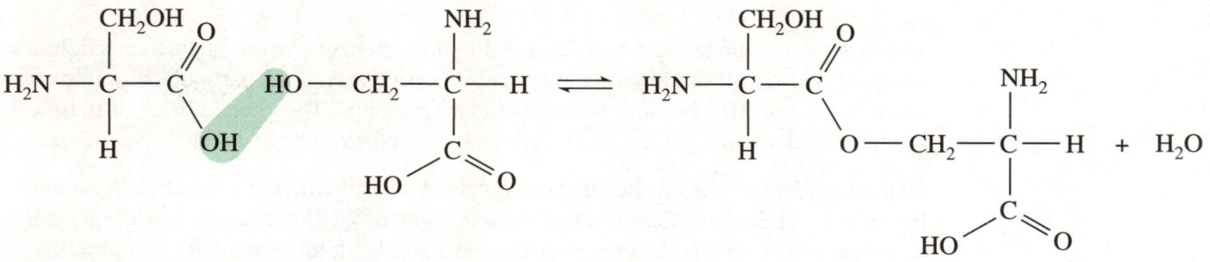

$$\underset{}{H_2N-\underset{H}{\overset{CH_2OH}{C}}-\underset{OH}{\overset{O}{C}}} \quad HO-CH_2-\underset{\underset{O}{\overset{}{C}}-OH}{\overset{NH_2}{C}}-H \quad \rightleftharpoons \quad H_2N-\underset{H}{\overset{CH_2OH}{C}}-\underset{}{\overset{O}{C}}-O-CH_2-\underset{\underset{O}{\overset{}{C}}-OH}{\overset{NH_2}{C}}-H \quad + \quad H_2O \qquad 39$$

serine ester dimer

The result is a twelve-membered ring (known as a triserine ring), with three NH$_2$ groups pointing away from the ring. These NH$_2$ groups are all pointing to the *same* side of the imaginary plane formed by the triserine ring (this is because natural serine exists as a single enantiomer). To complete the enterobactin structure, three catechol groups are attached to the NH$_2$ groups of the triserine ring via amide, C(O)—NH, linkages. The overall three-dimensional structure of enterobactin shows a triserine ring, to which three catechol groups are attached via the nitrogen atoms, all linked to the same side of the ring (Figure 68).

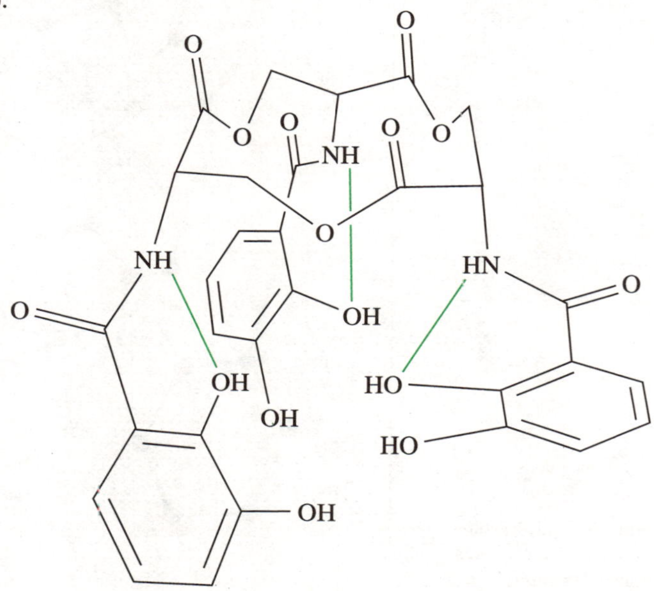

Figure 68 Three-dimensional structure of enterobactin.

Figure 68 also shows that there are other interactions. These interactions are three hydrogen bonds between the NH of a serine group and the oxygen atom of a catechol ring (shown as green lines in Figure 68). This imposes further structural rigidity by preventing each catechol ring from rotating freely, so that not only are the catechol groups all linked to the same side of the triserine ring, but all the catechol oxygen atoms face towards the centre of the ligand.

We see that the free enterobactin ligand is actually rather rigid in its structure, with all six coordinating oxygen atoms of the catechol groups held in position to bind an iron(III) ion (that is, the catechol groups have the same relative positions both *before* and *after* the binding of iron(III)). This rigid arrangement of functional groups before the metal has been chelated is known as **ligand preorganisation**; in other words, the three-dimensional structure of the ligand hardly changes on complex formation. What this means in practice is that a metal ion of a particular size (and charge) forms a particularly stable complex with the preorganised ligand. Iron(III) has the correct size and charge to form a very stable complex with the preorganised enterobactin ligand, and, therefore, is chelated selectively by enterobactin. This is exactly the same phenomenon as selective chelation of alkali metal ions by different sizes of crown ethers, as in Figure 8. A small crown ligand, like 12-crown-4 (reaction 40), will form stable complexes with a small metal ion, Li$^+$, whereas a larger crown, such as 18-crown-6, will form stable complexes with a larger metal ion, K$^+$ (reaction 41):

[Structures of 12-crown-4 + Li⁺ → [Li(12-crown-4)]⁺ complex (40), and 18-crown-6 + K⁺ → [K(18-crown-6)]⁺ complex (41).]

☐ Why does preorganisation of a ligand lead to a high stability constant?

■ This is a manifestation of the chelate effect. If ΔS for the reaction is positive, then the value of ΔG is more negative, and the stability constant is larger. In a reaction such as complexation of iron(III) with enterobactin, the hexadentate ligand hardly changes its structure, and this will have a larger entropy increase than a reaction to produce a hexacoordinated iron complex from six monodentate ligands. (Remember, however, that ΔS is an overall term for the reaction, and the ligand structure is only one part of it; it also takes into account the entropy effects due to the water molecules surrounding the metal and ligand, etc.)

SAQ 21 Summarise the features of enterobactin that make it selective for iron(III) and that make the iron(III)–enterobactin complex highly stable.

7.2.1 Removal of iron

Before leaving enterobactin to look at iron transport and storage in humans, it is worth asking the question: how does *E. coli* remove the iron from such a stable complex as the iron(III)–enterobactin once it has been absorbed?

The answer to this question can be found if we look back to reaction 38. The rigid, three-dimensional structure of the triserine ring of enterobactin is the main reason why enterobactin is such an effective ligand. If the structure of the ring is destroyed, enterobactin loses much of its chelating power. In fact, this is exactly what happens to the iron(III)–enterobactin complex once it has been absorbed by *E. coli*. Enzymes called **esterases** (so-named because they catalyse the decomposition and formation of esters) hydrolyse the triserine ring of enterobactin in the reverse reaction of reaction 38. This breaks up the triserine ring structure, and the stability constant of the resulting [Fe(catechol)₃] complex is much lower than 10^{49} mol⁻¹ l. Iron is then relatively easily removed from this complex, so that it can be used in the bacterium cell.

7.2.2 Summary of Section 7.2

1 *E. coli* has a remarkable method of obtaining iron from its environment, which involves the use of very powerful iron chelators, called siderophores.

2 One siderophore in particular, enterobactin, forms an extremely stable complex with iron(III).

3 The high stability of this complex is due partly to the rigid, preorganised structure of the ligand, and partly to the iron(III) being the correct size and charge to be chelated effectively by enterobactin.

4 Enzymes called esterases are able to catalyse the hydrolysis of the iron(III)–enterobactin complex and so release iron.

7.3 Iron transport and storage

As bacteria secrete such powerful chelators into the environment, iron in other organisms must be kept under very close control. Any free iron within an organism is likely to be chelated by a siderophore, which may lead to bacterial infection within the organism. In this Section we shall examine the biochemical systems that handle iron within the human body. The two areas we shall study are iron transport and iron storage.

7.3.1 Iron transport

It is obvious that iron must be transported around the human body. Firstly, it must be transported from the food in the gut to the places where it is required. Mostly, iron is required in the bone marrow, where red blood cells are formed. Red blood cells have a finite lifetime of about only four months, and old cells are destroyed, usually in the spleen. Iron from the destruction of these cells is then transported from the spleen back to the bone marrow to be recycled.

Iron cannot be transported around the body's circulation system as free iron, since it would be susceptible to chelation by siderophores, or may precipitate as iron(III) oxide, or may form iron(II). Therefore, a specific transport protein is required, called **transferrin**^{G}. (In fact, a whole class of transferrin-like proteins is involved in iron transport.) Transferrin is a medium-sized protein with a relative molecular mass of about 80 000. The crystal structures of transferrin with and without iron have been obtained, and the overall structure is shown in Figure 69. The structure without any iron (Figure 69a) shows that transferrin can be considered as two very similar polypeptides back-to-back, with each of the polypeptides having a large cleft. The apex of each cleft coordinates one iron atom; each transferrin molecule is therefore capable of transporting two iron atoms. Also, on the binding of iron, there is a significant change in the higher-order structure of the protein, such that the two sides of the cleft come together and incarcerate the iron atoms (Figure 69b). Both iron atoms are now buried deep within the protein structure. (It is not fully clear why the iron atoms are buried in this way, but it may help in protecting the iron atom from microbial siderophores.)

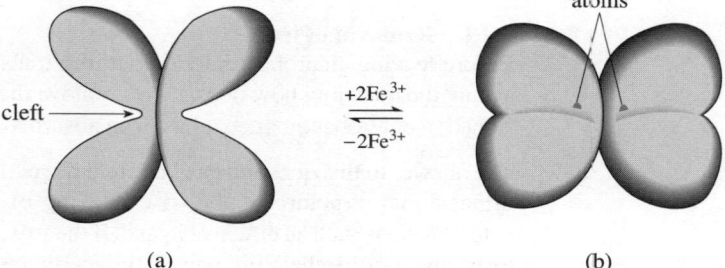

Figure 69 (a) Schematic diagram of transferrin protein; (b) proposed higher-order structure change on complexation of iron(III).

The iron binding site is rich in hard ligands, which are suitable for binding iron(III) in a stable complex (Figure 70a). When the iron atom enters the active site (Figure 70b) it is coordinated by one η^1-aspartyl, one histidyl and two tyrosinate side-chains; a non-protein ligand also coordinates to it. This external ligand is a carbonate, CO_3^{2-}, which is held in place within the protein via hydrogen bonds to the protein backbone.

☐ What is the mode of bonding of the CO_3^{2-} group?

■ The carbonate coordinates to the iron in an η^2 fashion; in other words, it is a chelating ligand.

Once the carbonate is held in place, we can see another example of ligand preorganisation, where an octahedral environment of hard/borderline ligands is ready to receive the iron(III) ion. The carbonate binding appears to facilitate the iron binding by the protein and vice versa, and so the system is said to be *synergistic*. It is not clear why this unusual synergistic binding of iron and carbonate occurs in transferrin, but it may have something to do with the way iron is released from transferrin.

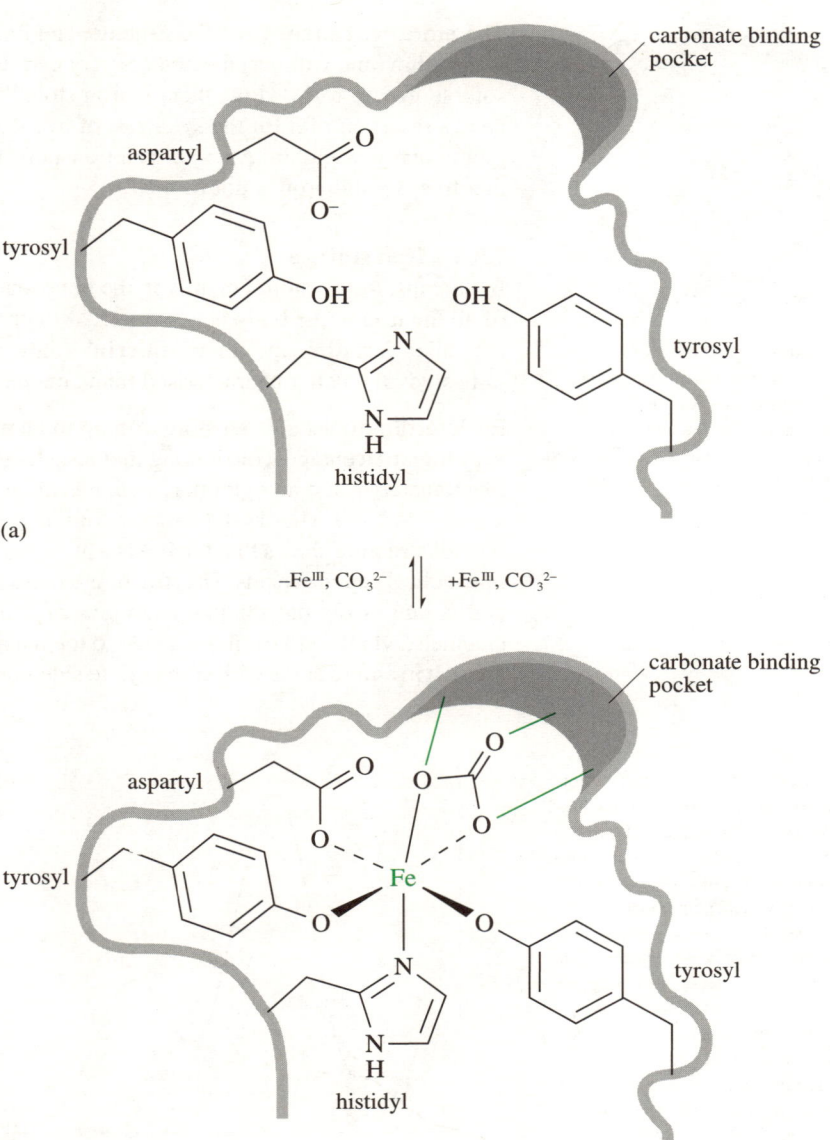

Figure 70 (a) The iron binding site in transferrin; (b) the six-coordinate iron site; the coordination geometry is distorted octahedral. The carbonate is held in place by hydrogen bonds (green lines) to amino acid side-chains inside a small cleft.

Since all the coordinating ligands in transferrin can be considered as hard or 'borderline' ligands, it is no surprise that transferrin forms a very stable complex with iron(III). The stability constant of the Fe(III)–transferrin complex is $c.\ 10^{20}$ mol^{-1} l. This is high enough to protect the iron(III) against the low concentration of any siderophores present. Indeed, the transferrins show mild antibacterial properties, in which their method of operation is to prevent extensive iron chelation by siderophores (see Box 3).

> ### Box 3 Iron in human milk
>
> It has been known for some time that bottle-fed babies are more likely to suffer from gastric infections than breast-fed babies; this may be despite strict hygiene standards. The reason for this probably lies in the availability of iron within the baby's feed. Breast milk is known to contain a transferrin-like protein called **lactoferrin**^{G}. The lactoferrin chelates all the iron in the mother's milk, and prevents iron chelation by microbial siderophores. Formula milk, on the other hand, does not contain human lactoferrin, so the iron in the feed is more available for chelation by siderophores secreted by bacteria.

Table 9 Stability constants of metal–transferrin complexes

Metal	$\log_{10}$ (stability constant)
cadmium(II)	5.95
zinc(II)	7.8
aluminium(III)	13.5
iron(III)	22.8

Transferrin also forms relatively stable complexes with other hard metals (Table 9). Current thinking suggests that these other metals are transported by transferrin into cells, where they are potentially detrimental. One element in particular, aluminium, is of concern because it is used widely in cooking utensils.

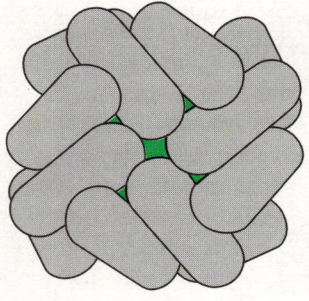

Figure 71 Sub-unit assembly of ferritin. Each sub-unit (shaped like a sausage) is made up of four parallel, α-helical polypeptide chains. The channel at the centre of the structure is clearly visible (dark green area); it lies on a fourfold axis of symmetry.

Therefore, we can see from the structure and function of transferrin, that the transport of iron within (and without) the body is very carefully managed, so as not to allow any free soluble iron to form. How, then, is iron stored? After all, we must store iron, since we need a reservoir of it for the synthesis of iron-containing proteins, most notably haemoglobin and myoglobin. As with iron transport, the iron storage systems need to ensure that free, soluble iron is not formed.

7.3.2 Iron storage

In humans, iron is stored mainly in the bone marrow, spleen and liver. About 10 per cent of all the iron in the body is in storage. Two proteins are involved in iron storage; these are called **ferritin**^{G} and **haemosiderin**^{G} (they also occur in other organisms). We shall only study the better characterised (and simpler!) ferritin.

Each ferritin molecule can store iron up to about 20 per cent of its total mass. This is a very high percentage, considering that less than 0.2 per cent of the total mass of proteins like transferrin and myoglobin is iron. Ferritin is a large protein with a relative molecular mass of 440 000. The crystal structure of ferritin with *no iron* is shown in Figure 71. The overall structure shows that ferritin is a huge, hollow protein, with a wall mostly made up of α-helical peptide chains. The structure is quite symmetrical, being roughly dodecahedral, and is one of the outstanding examples of symmetry in chemistry. The wall contains channels, which lead from the inside to the outside of the hollow 'sphere'. The channels are rich in amino acids with carboxylate side-chains, which are capable of chelating iron.

Figure 72 Fe-EXAFS radial distribution plot of iron-containing ferritin. Notice that there are two peaks, the first at 160 pm corresponding to a sphere of oxygen atoms, and the second at 290 pm corresponding to a sphere of iron atoms. (The peak due to the iron atoms is smaller than the peak due to the oxygen atoms; this is not in accord with the relative number of electrons in oxygen and iron atoms. The reasons for this are complex, but involve other factors beside the number of electrons in the intensity of back-scattering (see Section 4.2). Also the actual structure of the hydrated iron(III) oxide in ferritin is not 'perfect', in that there are incomplete 'shells' of iron atoms, and poor long-range crystal order.)

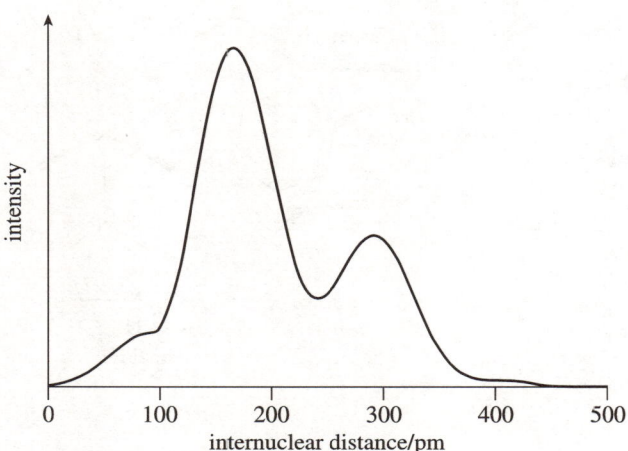

The crystal structure of iron-containing ferritin is not known. However, some clues as to its structure have been obtained from EXAFS studies. EXAFS, as discussed in Section 4, gives information about the direct coordination environment of a particular atom in terms of the number and type of its coordinated atoms (although no angular information is usually available). EXAFS studies on iron-containing ferritin showed that each iron atom is surrounded by an inner shell of six or seven oxygen atoms at a distance of about 160 pm, and by a second shell of seven or eight iron atoms at a distance of about 290 pm (Figure 72). This was a very strange result. How could seven or eight iron atoms be packed around each iron atom? The problem was solved when it was noticed that the EXAFS data were very similar to that of a hydrated iron(III) oxide mineral called ferrihydrite, $5Fe_2O_3.9H_2O$. From this result it was clear that ferritin stored iron partly as a crystalline, hydrated iron(III) oxide. Further studies showed that the inorganic crystalline part was within the hollow sphere of the protein (Figure 73).

Therefore, ferritin stores iron as crystalline, hydrated iron(III) oxide within its structure.

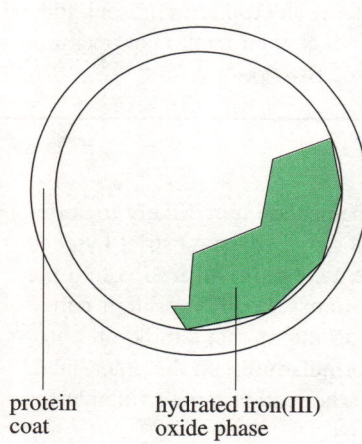

Figure 73 Schematic diagram showing growth of iron(III) oxide within a ferritin macromolecule. The full ferritin contains about 4 500 iron atoms.

☐   How will this affect the availability of the iron?

■   As the iron(III) oxide is very insoluble, it is unavailable to microbial iron-chelating ligands.

The inorganic iron(III) oxide core is also protected from chelators by the outer protein coat. Moreover, this is a very space-efficient method of storing iron; each ferritin protein macromolecule can store a maximum of 4 500 iron atoms.

Iron is delivered to ferritin (after having been transported by transferrin), where it migrates through the carboxylate-rich channels in the surface of the protein to the interior. The inner side of the protein sphere is also rich in carboxylate residues. It is thought that these carboxylate residues coordinate iron atoms, such that the iron atoms are held in a regular array. This regular array has the correct spacing of iron atoms to encourage the growth of crystals of iron(III) oxide, and in this way an iron(III) oxide phase grows within the ferritin core.

☐   What kind of process do we call this, and where have we met it before?

■   The growth of iron(III) oxide in ferritin is another example of biomineralisation. We saw exactly the same process in the growth of bone on collagen in Section 2.

So it is somewhat ironic that the formation of highly insoluble iron(III) oxides and hydroxides, which causes such a problem in the availability of iron in the environment, is the method by which iron is stored in ferritin!

7.4   Summary of Section 7

We have seen in this Section that, despite having a high natural abundance, iron is in very short supply because of the insolubility of its oxides and hydroxides. A result of this is that organisms have developed methods for the uptake, transport and storage of iron. Bacteria, in particular, secrete very powerful iron chelators known as siderophores. Of all the iron–siderophore complexes, the iron(III)–enterobactin complex has the exceptionally high stability constant of $10^{49}\,\text{mol}^{-1}\,\text{l}$.

Other organisms, partly as a defence against siderophores and the need to avoid free iron in solution, have biochemical methods to transport and store iron. The protein most associated with iron transport is transferrin. Iron storage, in mammals, including humans, is achieved by ferritin, which stores iron as a hydrated iron(III) oxide; this is an example of biomineralisation.

The key points of this Section are:

1   The concentration of free $Fe^{3+}(aq)$ at pH 7 is very small.

2   Free $Fe^{2+}(aq)$ in an organism is dangerous, because it can react to produce the superoxide radical anion.

3   Iron(III) tends to be coordinated by hard ligands such as those containing oxygen or nitrogen.

4   The chelate effect confers extra stability on iron(III) complexes with polydentate ligands.

5   Siderophores are powerful microbial iron chelators.

6   Enterobactin is a particularly powerful iron chelator. It acts as a preorganised hexadentate ligand for iron(III).

7   Esterases catalyse the hydrolysis of the triserine ring in enterobactin, thus enabling iron(III) from the enterobactin complex to be released.

8   Iron transport in mammals is carried out by transferrin, which holds iron(III) in a hexacoordinate site, one of the ligands being η^2-carbonate.

9   Ferritin is an iron storage protein capable of storing up to 4 500 iron atoms per protein macromolecule. The iron is stored as hydrated iron(III) oxide.

SAQ 22 What is a siderophore?

SAQ 23 Compounds **38** and **39** are synthetic siderophores. The stability constant of the iron–**38** complex is 10^{40} mol^{-1} l, whereas the stability constant of iron–**39** is 10^{28} mol^{-1} l. Explain why both stability constants are less than that for iron–enterobactin, and why the stability constant for iron–**38** is greater than that for iron–**39**.

SAQ 24 Structure **40** is the siderophore agrobactin. Sketch the conformation you would expect it to adopt in the complex it forms with iron(III), and indicate the location of the iron(III) binding site.

---

8 ZINC: A CASE STUDY

So far we have concentrated on the roles of iron as a bioinorganic element. Iron is the most important trace element in biochemical systems. In this Section, we shall briefly examine zinc in the context of bioinorganic chemistry. A look back at Table 4 in Section 3 shows us that zinc is an important bioinorganic element also. Indeed, the human adult's recommended daily intake of zinc is almost the same as that of iron, and zinc occurs in a wide range of enzymes and proteins. Why is zinc such an important bioinorganic element? To answer this, we first need to examine the chemistry of zinc.

8.1 Zinc chemistry

In the Periodic Table, zinc is sandwiched between the first-row transition metals and Groups III–VII of the fourth-Period main-Group elements.

☐ Is zinc classed as a transition element?

■ Not usually. The criterion for a transition element is that the atom has a partially filled d sub-shell. The only important oxidation state of zinc is zinc(II), which has the electronic configuration [Ar]$4s^2 3d^{10}$.

So zinc compounds exhibit one major oxidation state, +2. Apart from zero, other oxidation states can be ignored for the purposes of this Section. The +2 state is stable, since in this oxidation state the zinc(II) ion has a full ten electrons in its 3d orbitals. In this respect, zinc differs from the first-row transition elements, in so far as it can be thought of as essentially redox inactive; that is, its oxidation state doesn't change during reactions. Therefore, we expect its bioinorganic chemistry to be very different from that of iron.

☐ What is the crystal-field stabilisation energy for zinc(II) in tetrahedral coordination geometry?

■ Zinc(II) has no crystal-field stabilisation energy in either the tetrahedral or octahedral case, since its d orbitals are fully occupied. Hence neither geometry is strongly favoured on energetic grounds.

In practice, the coordination complexes of zinc are found mostly to have four-coordinate tetrahedral geometry, although some five- and six-coordinate complexes are known. Zinc(II) is classed as a 'borderline' acid in hard–soft acid–base theory, and, as such, will form stable complexes with a variety of ligand types.

☐ Would you expect zinc(II) complexes to be either coloured or magnetic?

■ With a configuration of $3d^{10}$, we would expect the complexes normally to be colourless and diamagnetic. If any zinc(II) compound is coloured (for example, zinc(II) selenide, ZnSe, is orange), the colour comes from charge-transfer or ligand-to-ligand electronic transitions, since no d–d transitions occur in the fully occupied d^{10} configuration.

Since most zinc complexes are colourless and diamagnetic, they are referred to as **spectroscopically silent**, and spectroscopic investigation of zinc(II) complexes does not give much information.

Zinc(II) is a reasonably strong Lewis acid. In organic synthesis, zinc salts are often used as catalysts, where the zinc acts as a Lewis acid. For instance, in a Friedel–Crafts alkylation of a benzene ring, $ZnCl_2$ is sometimes used as a catalyst:

$$\text{PhR} + \text{R'Cl} \xrightarrow{ZnCl_2} \text{R-C}_6\text{H}_4\text{-R'} + \text{HCl} \qquad 42$$

The mechanism of this reaction is shown as steps 43–45; the zinc(II) polarises the attacking substrate, and thereby enhances its electrophilicity:

$$R'\!-\!Cl + ZnCl_2 \rightleftharpoons R'^{\delta+}\!-\!Cl \rightarrow ZnCl_2^{\delta-} \qquad 43$$

$$44$$

$$45$$

Zinc(II) will also polarise coordinated water molecules, and zinc(II) salts dissolved in water will undergo hydrolysis, in a similar fashion to iron(III):

$$Zn^{2+}(aq) + H_2O(l) = [Zn(OH)]^+(aq) + H^+(aq) \qquad 46$$

$$[Zn(OH)]^+(aq) + H_2O(l) = Zn(OH)_2(s) + H^+(aq) \qquad 47$$

At first glance, zinc(II), with its redox-inactive behaviour, appears to be an uninteresting element; despite this, it is found in a very wide range of essential proteins and enzymes. Indeed, its rather unreactive behaviour and polarising ability are what make it useful in biological systems, where it can be used to interact with substrates without undergoing any redox reaction.

From the zinc proteins studied so far, it has been discovered that 'biological' zinc(II) falls into two classes, these are: (i) where zinc acts as a Lewis acid, and (ii) where zinc acts as a structural element of the protein. We shall examine examples of both roles.

8.2 Zinc(II): biology's Lewis acid

8.2.1 Carbonic anhydrase

The first enzyme we shall look at where zinc(II) behaves as a Lewis acid is **carbonic anhydrase (CA)**[G]. CA is an enzyme that is present in many organisms, including humans, where it is found in the erythrocytes (red blood cells). It was first discovered in 1932, and in 1939 it was shown that CA contained zinc. The enzyme catalyses the following physiologically important reaction, where carbon dioxide and water react to make hydrogen carbonate, HCO_3^- (commonly referred to as bicarbonate):

$$CO_2(g) + H_2O(l) = HCO_3^-(aq) + H^+(aq) \qquad 48$$

This is an important reaction because CO_2 is released by our cells as a by-product of respiration. CO_2 is a gas at body temperature with only a limited solubility in water, and therefore it cannot be easily transported in the bloodstream. However, converting CO_2 to hydrogen carbonate, which is very soluble in water, effectively increases greatly the solubility of carbon dioxide in water. But, if our cells respire at a rapid rate, the above reaction must occur very quickly indeed if the rate of respiration by excessive gaseous CO_2 build up in the blood is not to be limited.

At pH 7, which is roughly the pH of blood serum, the uncatalysed forward reaction 48 occurs at a rate of about $0.1\,\text{mol}^{-1}\,\text{l}\,\text{s}^{-1}$; this is far too slow a conversion rate to convert all the CO_2 generated by respiration into hydrogen carbonate.

☐ At pH 9 the forward reaction in equation 48 is much greater, at $10^4\,\text{mol}^{-1}\,\text{l}\,\text{s}^{-1}$. Why?

■ At pH 9 there is a much higher concentration of hydroxide ions in solution than at pH 7 (roughly 100 times higher). In reaction 48, the first step is the nucleophilic attack of either water or hydroxide on the CO_2:

$$H_2\ddot{O} + CO_2 \longrightarrow H-\overset{+}{O}(H)-C(=O)(O^-) \longrightarrow HCO_3^- + H^+ \qquad 49$$

$$HO^- + CO_2 \longrightarrow HCO_3^- \qquad 50$$

Formation of hydrogen carbonate is much quicker if the nucleophile is hydroxide ion (reaction 50) than if it is water (reaction 49). Hence, at pH 9, the reaction is much quicker than at pH 7.

Since the pH of blood cannot be changed from *c.* pH 7, reaction 48 needs to be catalysed if it can be used as part of a CO_2 transport system in the body. Indeed, CA catalyses the reaction, such that the forward reaction now proceeds at $10^8\,\text{mol}^{-1}\,\text{l}\,\text{s}^{-1}$ (near

Figure 74 Schematic diagram of the higher-order structure of carbonic anhydrase; see also Plate 3 (inside back cover).

the diffusion-controlled limit in solution of about $10^9 \, mol^{-1} \, l \, s^{-1}$), giving an enormous 10^9 rate enhancement over the non-catalysed reaction!

What are the structural and chemical features of CA that make it such an efficient catalyst? CA is a small/medium protein with a relative molecular mass of $c.$ 30 000. It has a roughly spherical shape, with a large cleft in the side of the sphere. The cleft contains the active site and is about 1 500 pm deep (Figure 74). At the bottom of the cleft lies a zinc(II) ion coordinated by three histidyl side-chains (Figure 75). The zinc also has a water molecule (or hydroxide; see later) coordinated to it. Adjacent to the zinc in the active site is a group of hydrophobic amino acid residues, which forms a hydrophobic 'pocket'.

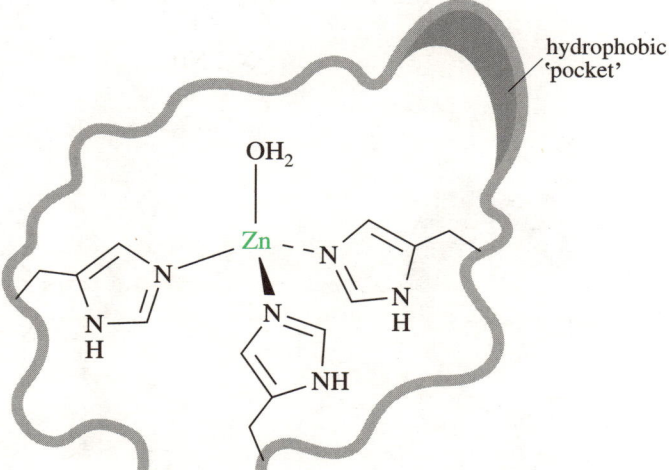

Figure 75 Close up of zinc(II) active site in carbonic anhydrase.

The key to the catalytic action of CA lies in its ability to generate a reactive hydroxide ion at pH 7. Reaction 51 shows that the water molecule coordinated to the zinc ion readily loses a proton to give a hydroxide ligand:

$$\text{[Zn(OH}_2\text{)(His)}_3\text{]} \rightleftharpoons \text{[Zn(OH)(His)}_3\text{]}^- + H^+ \qquad 51$$

This occurs easily because the zinc is a Lewis acid. In effect, the Lewis acidity of the zinc is translated into Brønsted acidity of the coordinated water.

☐ What is the difference between a Lewis acid and a Brønsted acid?

■ A Brønsted acid is a proton donor, whereas a Lewis acid is defined as an electron pair acceptor.

The water coordinated to the zinc is actually considerably more acidic than normal uncoordinated water. The pK_a of the zinc-bound water is 7.4, whereas normal water has a pK_a of 15.7 (at 25 °C).

The other reactant, CO_2, is bound in the active site next to the hydroxide ligand by the hydrophobic pocket of amino acids. Since the CO_2 and nucleophilic hydroxide are brought in close proximity with each other, they react together very rapidly to give the product hydrogen carbonate. The whole catalytic cycle of CA is shown in Figure 76.

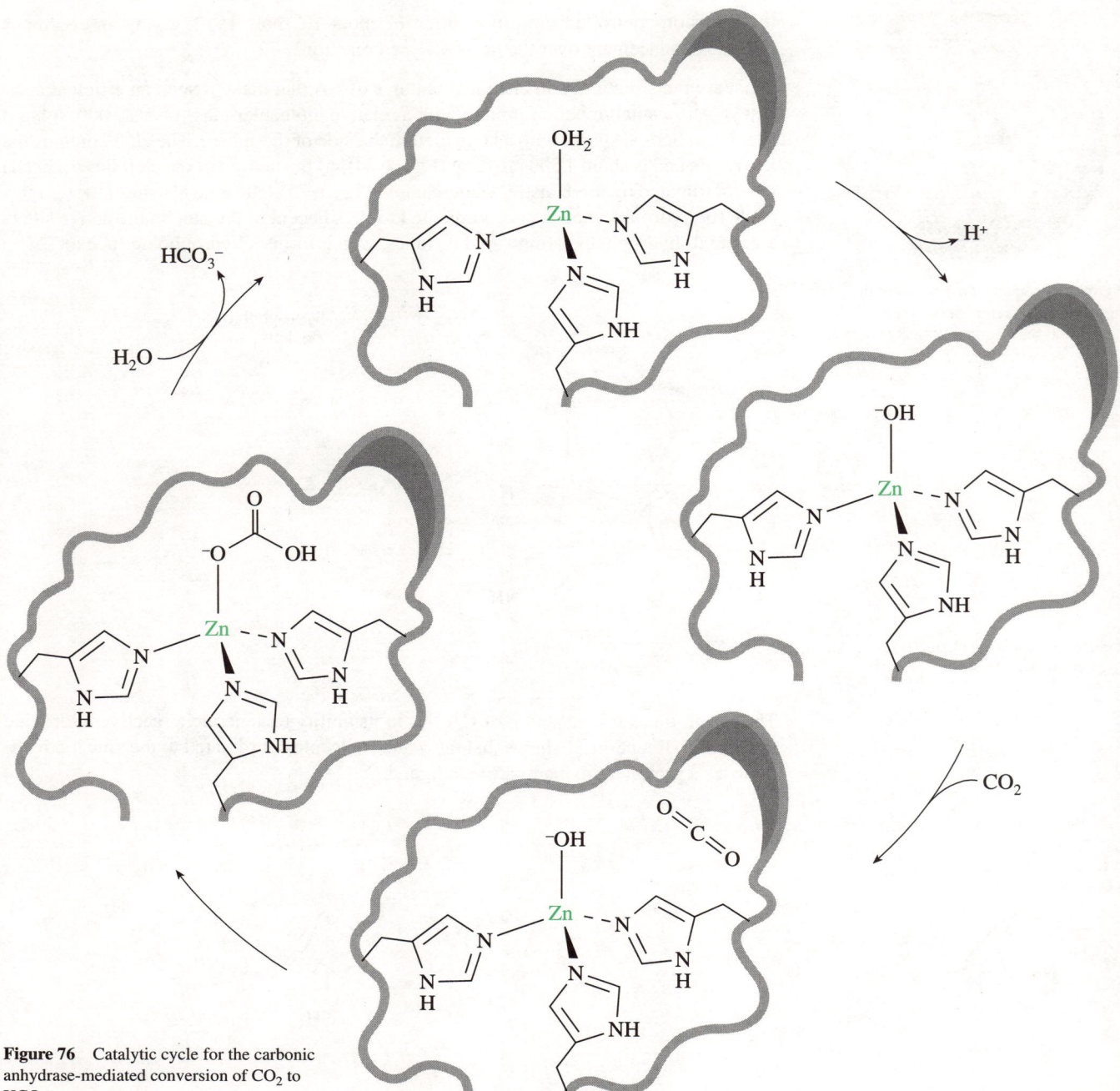

Figure 76 Catalytic cycle for the carbonic anhydrase-mediated conversion of CO_2 to HCO_3^-.

In CA, zinc(II) acts as a Lewis acid in the deprotonation of H_2O. For this reaction there is no need for a change in the oxidation state of the metal.

8.2.2 Liver alcohol dehydrogenase

Another enzyme that contains zinc is **liver alcohol dehydrogenase (LADH)**, the function of which is to catalyse the oxidation of primary alcohols to aldehydes (or secondary alcohols to ketones). For example:

$$CH_3CH_2OH = CH_3CHO + H^+ + H^- \qquad 52$$
$$\text{ethanol} \qquad \text{ethanal}$$

The products of this reaction include a hydride ion, H^-, which is normally very strongly reducing, and is potentially detrimental to the cell. During the LADH catalytic cycle the hydride is not generated directly, but is transferred to a co-substrate called nicotinamide adenine dinucleotide (abbreviated to NAD^+ [G] in its oxidised form and to NADH in the reduced form; it is a derivative of the important biochemical molecule adenosine 5-triphosphate, ATP[G]). The structure of nicotinamide is shown in Figure 77.

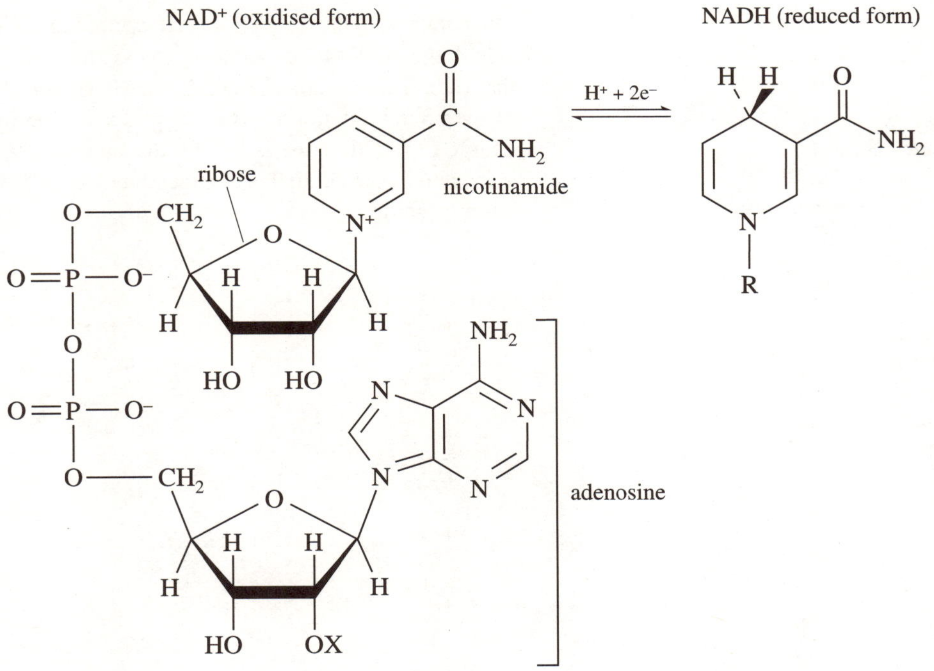

Figure 77 Structures of NAD$^+$ and NADH (X is H in NAD$^+$, and PO$_3^{2-}$ in NADP$^+$; R = ribose–diphosphate–adenosine).

The overall reaction catalysed by LADH is as follows (where ethanol is used as an example):

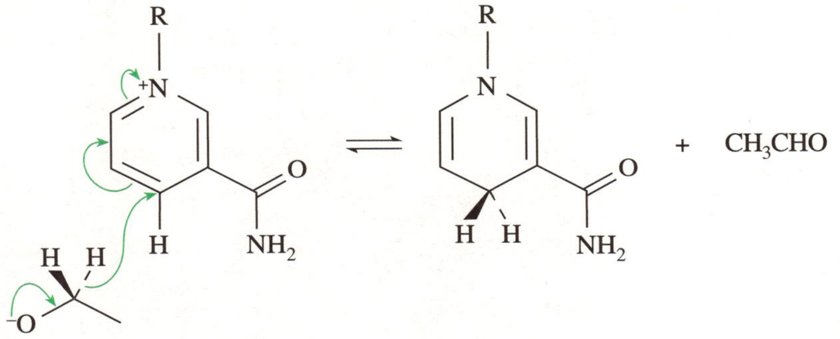

$$\text{(structure with arrows)} \rightleftharpoons \text{(reduced structure)} + CH_3CHO \qquad 53$$

or more simply as

$$CH_3CH_2OH + NAD^+ = CH_3CHO + H^+ + NADH \qquad 54$$

(LADH is obviously important in the metabolism of alcohol in the body and therefore is the centre of much research.)

The crystal structure of LADH has been determined, and has been found to contain two types of zinc(II) ion. In one of these, the zinc is coordinated by four cysteinyl side-chains (Figure 78a), and has no immediate role in the catalytic action of the enzyme. It is in fact playing a structural role, which we shall examine later. The other zinc ion is coordinated by two cysteinyl side-chains, one histidyl side-chain and one water molecule; it is this zinc that is essential for catalytic activity (Figure 78b).

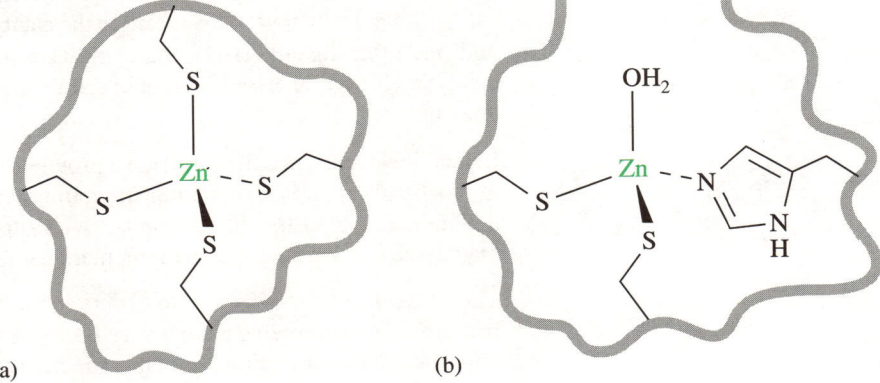

Figure 78 Close-up of the two zinc sites in liver alcohol dehydrogenase: (a) the zinc with a structural role (the cysteinyl residues are shown as 'S'); (b) the zinc that takes part in the catalytic cycle.

The catalytic cycle of LADH is shown in Figure 79. It shows some important features. In step **A** the NAD⁺ binds adjacent to the zinc site. In step **B** the alcohol substrate displaces the water from the zinc. In step **C**, the Lewis acidity of the zinc facilitates the deprotonation of the bound alcohol to give a bound alkoxide (analogous to H_2O in CA). The negative charge of the alkoxide is lost by the rapid transfer of H^- to NAD⁺ (step **D**). In the final step, step **E**, the ethanal is displaced by water, NADH leaves the active site, and the cycle begins again.

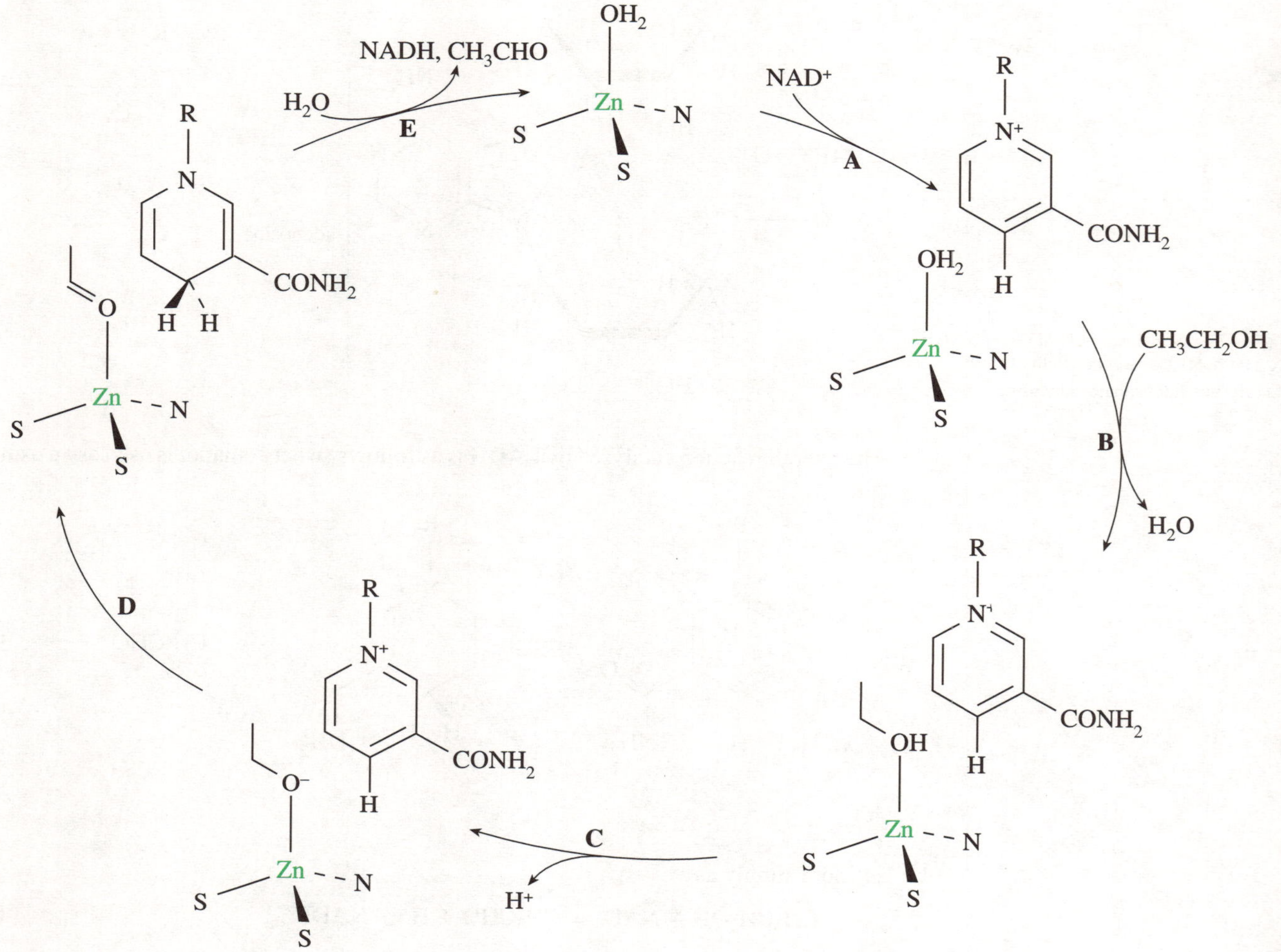

Figure 79 Catalytic cycle of liver alcohol dehydrogenase, using ethanol as an example; the cysteinyl residues are represented as 'S' and the histidyl residue as 'N'.

As in CA, the zinc acts as a Lewis acid and facilitates the deprotonation of the substrate, which in this case, is an alcohol.

8.3 Zinc(II) in a structural role

The structure of LADH shows two different types of environment around the zinc atoms. The zinc atom necessary for catalysis has been described. The other zinc is not necessary for catalysis. Indeed, it is unlikely that any catalysis occurs at this site because this zinc is already coordinated by four cysteinyl side-chains, so would not be expected to bind a further ligand/substrate. This is unlike the catalytic zinc, which has three protein ligands and one other ligand (H_2O); the cycle also involves NAD⁺ as a substrate. The non-catalytic zinc has an essential role in stabilising the overall higher-order structure of the enzyme.

In fact, the occurrence of zinc(II) in a protein's structure is now known to be widespread in biochemistry; it is an essential structural element of many proteins. We have already met an example of this in Section 6.3 with superoxide dismutase, where the zinc ion's major role in the active site is to maintain its three-dimensional structure.

This structural type of zinc also occurs in another class of proteins called **zinc-finger proteins**. These proteins have a very precise and rigid higher-order structure, which is important in their function. A notable function of zinc-finger proteins is to recognise and bind to DNA. The higher-order structure of DNA involves a double helix, which is important in its function. The zinc-finger proteins work by wrapping around the DNA double helix; they have 'finger-like' higher-order structures, which fit inside the grooves of the helix (Figure 80). The importance of this is that the proteins can recognise specific parts (base sequences[G]) of DNA, because of the rigid higher-order structure.

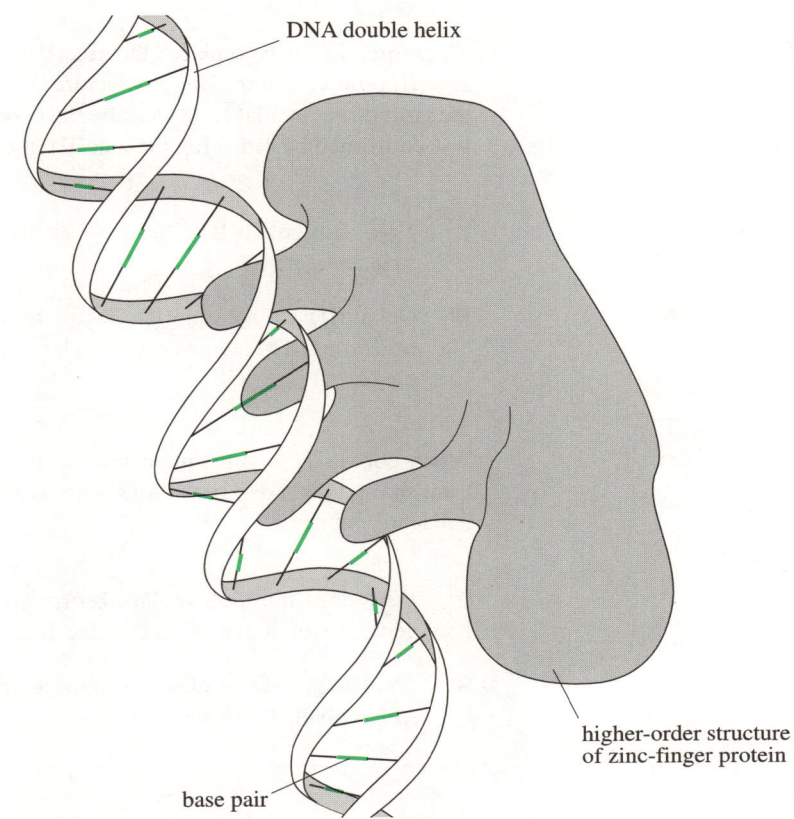

Figure 80 Schematic diagram of zinc-finger protein interacting with the double-helix structure of DNA.

Removal of zinc from zinc-finger proteins inactivates them for DNA-binding. Single-crystal X-ray crystallography of a zinc-finger protein reveals that zinc is present in the proteins as a structural element. In fact, zinc is needed to maintain the finger-like structure of the protein. Figure 81 shows how this is achieved. The zinc ion is actually at the 'base' of the 'finger', where it is coordinated by four amino acid side-chains: two histidyls and two cysteinyls. The rest of the finger is a loop of polypeptide chain. If the zinc were not present, the finger shape would no longer be pinned at its base; it would then collapse, with subsequent loss of protein activity. This was confirmed by an important experiment, in which the zinc-free protein was reactivated by the addition of a cobalt(II) salt (Figure 82).

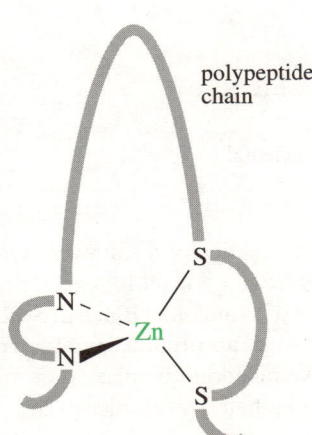

Figure 81 Schematic representation of the zinc-finger structure; N represents a histidyl side-chain, and S represents a cysteinyl side-chain.

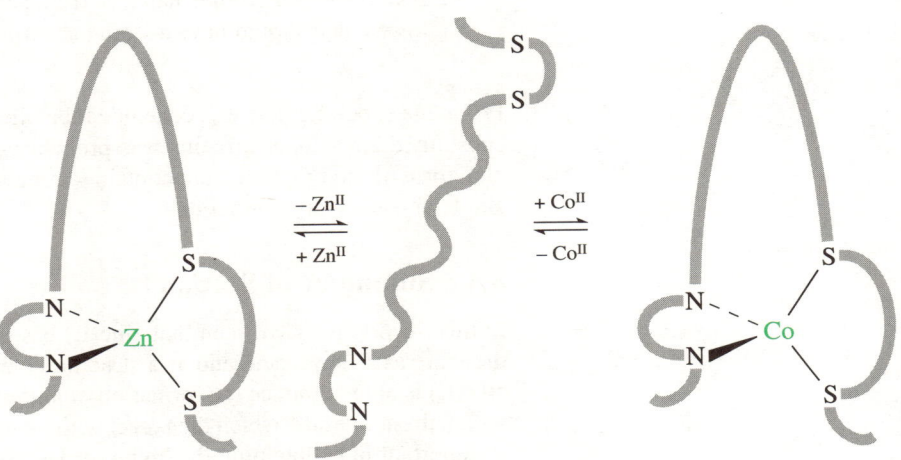

Figure 82 Schematic representation showing how zinc(II) is an essential structural feature of zinc-finger proteins, and how the finger structure can be reconstructed with cobalt(II).

☐ Zinc(II) readily forms tetrahedral complexes. Cobalt(II) is a good replacement for zinc(II) because the d^7 configuration of cobalt(II) gives relatively stable tetrahedral complexes. Calculate the CFSE of tetrahedral iron d^6, cobalt d^7 and nickel d^8 complexes to show that d^7 is the most stable.

■ The CFSEs are: iron(II), $0.6\Delta_t$; cobalt(II), $1.2\Delta_t$; nickel(II), $0.8\Delta_t$.

Therefore, cobalt(II) replaces the zinc(II) ion, forming a tetrahedral complex, as in the zinc(II) protein, and the finger structure is reformed. In fact, before a crystal structure of the protein was available, the tetrahedral coordination geometry of zinc(II) in this protein was confirmed by replacing the zinc(II) in the protein with cobalt(II).

☐ How does cobalt(II) differ from zinc(II), and therefore what method might we use to investigate it?

■ Cobalt(II), unlike zinc(II), has partially filled d orbitals (d^7), which exhibit d–d transitions. These transitions can be observed in the visible spectrum of Co(II) complexes.

When cobalt(II) is substituted into a zinc-finger protein, the resulting visible spectrum shows cobalt d–d transitions with large molar absorption coefficients (greater than $500\,l\,mol^{-1}\,cm^{-1}$).

☐ Why does the high molar absorption coefficient suggest that the coordination geometry of cobalt(II) in the zinc-finger protein is tetrahedral and not octahedral?

■ The two possible d-electron splitting diagrams for tetrahedral and octahedral (high spin) cobalt(II) are shown below.

tetrahedral   octahedral

In the octahedral complex, any d–d transition is theoretically disallowed by the symmetry selection rule, which forbids transitions from a g level to a g level and from a u level to a u level in complexes with a centre of symmetry (Block 2, Section 8.1). Therefore we would only expect to see a very weak absorption band for a d–d transition. On the other hand, d–d transitions in tetrahedral complexes are fully allowed, and should have a strong absorption band in their u.v./visible spectra.

From this experiment, it was concluded that the coordination environment in cobalt(II)-substituted zinc-finger proteins was probably tetrahedral, and, by analogy, the natural zinc form of the protein would contain tetrahedral zinc(II). This was later confirmed by single-crystal X-ray diffraction.

8.4 Summary of Section 8

In this Section we have seen that zinc(II) has two major roles in biochemical systems: these are as a Lewis acid and as a structural element of proteins. In its Lewis acid role, zinc(II) is able to induce the formation of anionic ligands (like hydroxide and alkoxide) within the active site, which then react with other substrates. In its structural role, zinc(II) is important in maintaining the higher-order structure of several proteins, including the structure of the important zinc-finger proteins.

The main points from this Section are:

1 Zinc is an important bioinorganic element.

2 The fact that zinc(II) has a $3d^{10}$ configuration makes it difficult to study by spectroscopic or magnetic means, and much of what is known about zinc in bioinorganic chemistry comes from crystal structures of zinc enzymes and proteins.

3 The redox-inactive behaviour of zinc(II) makes it useful in bioinorganic chemistry as (i) a Lewis acid and (ii) a structural element.

4 Zinc appears in a wide range of proteins and enzymes: carbonic anhydrase catalyses the solubilisation of carbon dioxide; liver alcohol dehydrogenase catalyses the breakdown of alcohols; zinc-finger proteins wrap around DNA.

SAQ 25 What other elements could substitute for zinc(II) in its biological role?

SAQ 26 The zinc(II) ion in carbonic anhydrase can be substituted by first-row transition metal ions, and the activity of the metal-substituted enzyme can be measured. The following results were obtained for the ability of metal-substituted enzymes to catalyse the reaction:

$$Zn^{II} > Co^{II} > Ni^{II} \approx Cu^{II}$$

Explain the trend in reactivity.

9 THE FUTURE OF BIOINORGANIC CHEMISTRY

In this Block, we have seen the widespread occurrence of inorganic elements in biochemical systems. Some of these elements are of critical importance, such as iron and zinc. Iron in particular has an extensive bioinorganic chemistry, ranging from O_2 transport to mono-oxygenases (cytochrome P450). We have also seen that the biological chemistry of iron is closely associated with its inorganic chemistry, such as its common coordination numbers, possible oxidation states, etc. We also saw that the simple inorganic chemistry of zinc is reflected in its chemistry in biochemical systems.

We have also seen how physical techniques, notably resonance Raman spectroscopy and EXAFS, are quite specialised in bioinorganic chemistry: they have proved crucial in studying inorganic elements in proteins. The value of most of these techniques depends on the fact that they only give information about the particular metal and its immediate surroundings, while ignoring the bulk (and complexity!) of the rest of the organic structure. Without these techniques, bioinorganic chemistry could never have developed as rapidly as it has over the past few years.

Where does the future of bioinorganic chemistry lie? Certainly, many metalloproteins remain to be discovered and characterised, and their study will continue. In particular, a rapidly growing area is the use of inorganic compounds in medicine. Some inorganic compounds are already available as drugs/medicines. Maybe the most familiar are antacids for indigestion, which are usually solid bases like calcium carbonate. Less familiar are drugs like lithium carbonate, which is used to control psychological illnesses, and *cis*-platin[G] (*cis*-diamminedichloroplatinum(II), structure **41**), which is a powerful anti-cancer drug. Table 10 lists some of the current inorganic drugs/medicines and their therapeutic uses.

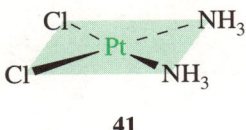

41

$[Pt^{II}Cl_2(NH_3)_2]$, or *cis*-platin, is a simple platinum compound that was rather fortuitously found to have anti-tumour properties. Its mode of action is interesting, and builds on some of the material that we have encountered in this Block.

Table 10 Some of the inorganic drugs, medicines and health products currently available for human use

Compound	Use/effect
cis-platin and other related compounds (Pt compounds)	anti-cancer
auranofin and other related compounds (Au compounds)	anti-arthritic
$[Au(Ph_2P(CH_2)_2PPh_2)_2]Cl$	anti-cancer
calcium carbonate	antacid
bismuth(III) citrate	anti-ulcer
$Na_2[Fe(NO)(CN)_5]$	for high blood pressure (vasodilator)
lithium carbonate	for manic depression
gadolinium(III) compounds	contrast agents for magnetic resonance imaging
barium sulphate	X-ray imaging agent
magnesium oxide	antacid
magnesium sulphate	laxative
aluminium hydroxide	antacid
fluoride	anti-tooth decay
silica	abrasive in toothpaste
tin compounds	for jaundice and anti-cancer
technetium compounds	myocardial and neurological imaging agents
sodium carbonate and sodium hydrogen carbonate	antacid and laxative
zirconium compounds	antiperspirant
titanium compounds	anti-cancer
strontium compounds	anti-plaque
zinc compounds	anti-plaque
many elements	dietary supplement

In the first place, *cis*-platin is a neutral compound. Since it is neutral, it can pass relatively easily across the lipophilic membrane that surrounds many cells. In this respect *cis*-platin differs from metal ions, which need to be chelated by a ligand before they can pass through a cell membrane (see Section 7.2 on siderophores). Once inside the cell, the *cis*-platin experiences a much lower chloride concentration than outside the cell (Figure 83). As a result and according to Le Chatelier's principle, the *cis*-platin's chloride ligands are substituted by water. The resulting charged (+2) complex cannot pass back across the cell membrane easily: it is trapped in the cell. Compare and contrast this method of metal uptake with that of iron by bacteria.

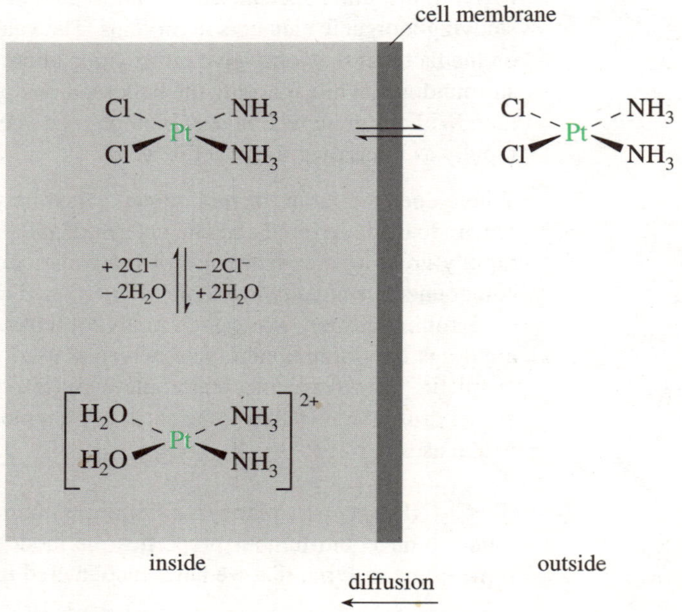

Figure 83 Schematic diagram of the uptake of *cis*-platin into a cell. The neutral *cis*-platin complex crosses the cell membrane via passive diffusion. Inside the cell, where the chloride concentration is much less than outside the cell, the *cis*-platin is hydrolysed to give a charged diamminediaquo complex.

Secondly, as platinum(II) is a borderline acid in terms of hard/soft acid–base theory, it forms relatively stable complexes with borderline bases. These bases can be found throughout the cell. However, it appears that *cis*-platin derivatives form particularly stable complexes with nucleobases, especially guanosine (structure **42**). Thus, when a *cis*-diamminoplatinum complex is coordinated by the imidazole nitrogen atoms of two guanosines on opposite strands of DNA, the DNA double-helix structure is disrupted (Figure 84). Since the double-helical structure of DNA is essential for its activity (such as recognition of shape by zinc-finger proteins, Section 8), a DNA molecule whose structure has been disrupted cannot function properly. Indeed, DNA plays a central role in cell replication, in which a copy is made of the DNA for a new cell. *Cis*-diamminoplatinum-coordinated DNA cannot be replicated, due to the deformation in its shape, and the result is cell death. For some unknown reason, *cis*-platin is effective at killing certain types of cancer cells outright, whereas healthy cells are able to recover from the effects of *cis*-platin. The result is a drug that is extremely effective at curing testicular and ovarian cancers.

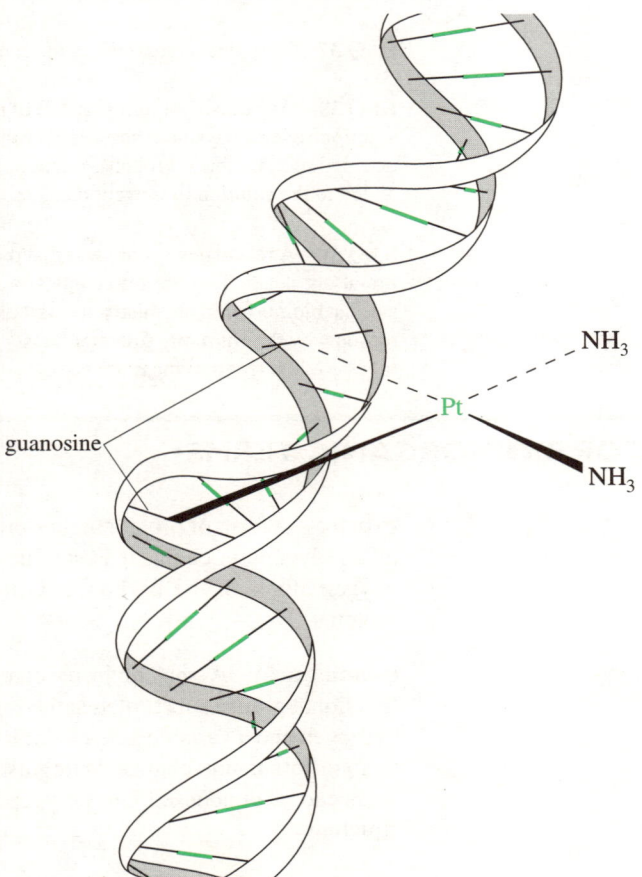

42

Figure 84 Sketch of DNA structure disruption by a *cis*-diamminoplatinum complex. The DNA is effectively bent, along with a slight unwinding of the helix.

The example of *cis*-platin shows how knowledge of metal uptake by cells, and the chemistry of the metal compound once inside a biochemical system, are both very important in the development of new drugs and medicines based on inorganic compounds. It is knowledge of, and research into, this kind of system that will be so important in the future.

You may now like to try the following questions to test your understanding of some of the concepts and material in this Block.

SAQ 27 If you were to design a molecule to bind oxygen reversibly, what structural and chemical features would you include?

SAQ 28 In aqueous solution, at pH 7, plutonium exists mostly as plutonium(IV). The charge : ionic radius ratio of Pu^{4+} is very similar to that of Fe^{3+}. How do you expect plutonium to be taken up in the body? Design a ligand that might be used to treat cases of plutonium poisoning. What side effects might it have?

SAQ 29 Sketch the radial distribution curve, up to about 250 pm, that you would expect to obtain from an Fe-EXAFS experiment for the haem group in myoglobin (see Figures 12 and 34).

SAQ 30 Why is zinc unsuitable as a substitute for iron in myoglobin?

SAQ 31 Describe some of the effects of too little and too much fluoride in the diet.

SAQ 32 Calculate the amount of free $Fe^{3+}(aq)$ in one litre of a $1\ mol\ l^{-1}$ solution of iron–enterobactin$^{(3-n)+}$ (ignore solubility effects and assume that the complex is fully formed).

SAQ 33 Without referring to Figure 61, sketch the catalytic cycle of superoxide dismutase. What is the role of each metal ion at the active site?

SAQ 34 Why are the cyanide ion and carbon monoxide very toxic?

SAQ 35 Do you expect oxygenated myoglobin to be diamagnetic or paramagnetic? Explain your reasoning.

SAQ 36 Why are both copper and iron suitable as O_2-binding sites in proteins that transport O_2?

SAQ 37 Why does transferrin act as a mild antibiotic?

SAQ 38 An aqueous solution (pH 7) of iron(II)–edta and DNA was exposed to air. Analysis of the solution several days later showed that the DNA had been chemically damaged (that is, the DNA double helix had been broken). Account for this observation, and suggest what might have been added to the solution to prevent damage.

SAQ 39 A recent news scare was that the fire retardants in cot mattresses were a possible cause of infant cot death. The fire retardants are often made from antimony compounds. Antimony has a similar bioinorganic chemistry to mercury, and both are toxic in high concentrations. Suggest a reason why the antimony fire retardants were considered dangerous, even though they could not be ingested directly from the mattress.

GLOSSARY OF BIOINORGANIC TERMS*

α-helix A form of protein higher-order structure, involving a helical shape for a section of a polypeptide chain. It is stabilised by hydrogen-bonding between NH groups and oxygen atoms of the backbone. Often represented as a cylinder-shape in diagrams of proteins.

α-amino acid A small molecule containing amine, $-NH_2$, and carboxylic acid, $-COOH$, functional groups substituted at the same (α) carbon atom. This carbon atom also has one hydrogen and a variable side-chain, R, attached to it: $NH_2CHRCOOH$ (Table 11). The 20 or so different side-chains, distinguish different naturally occurring amino acids. Amino acids can be condensed into polypeptide chains, which make-up the bulk of a protein's structure.

* References to terms defined elsewhere in this Glossary are printed in italics.

Table 11 The naturally occurring α-amino acids; the majority of amino acids have the general formula H₂N—CHR—COOH; the side-chain R may have any of the structures shown here

Amino acid	Symbol	R group
alanine	Ala	CH_3—
arginine	Arg	$H_2N-C(=NH)-NH-(CH_2)_3-$
asparagine	Asn	$CONH_2-CH_2-$
aspartic acid	Asp	$COOH-CH_2-$
cysteine	Cys	$SH-CH_2-$
glutamic acid	Glu	$COOH-(CH_2)_2-$
glutamine	Gln	$CONH_2-(CH_2)_2-$
glycine	Gly	$H-$
histidine	His	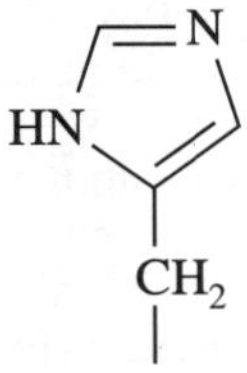 imidazole-CH_2-
isoleucine	Ile	$H_3C-CH(CH_2CH_3)-$
leucine	Leu	$(H_3C)_2CH-CH_2-$

Amino acid	Symbol	R group
lysine	Lys	$NH_2-(CH_2)_4-$
methionine	Met	$CH_3-S-(CH_2)_2-$
phenylalanine	Phe	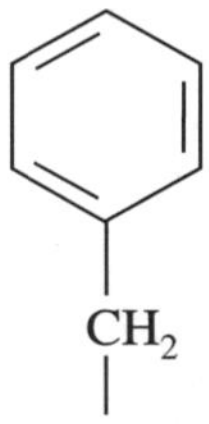 $C_6H_5-CH_2-$
proline (whole structure)	Pro	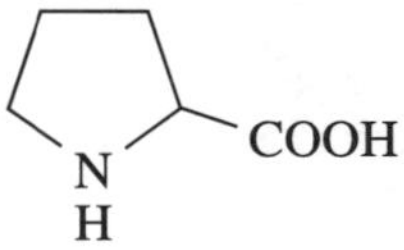 pyrrolidine-2-COOH
serine	Ser	$OH-CH_2-$
threonine	Thr	$OH-CH(CH_3)-$
tryptophan	Trp	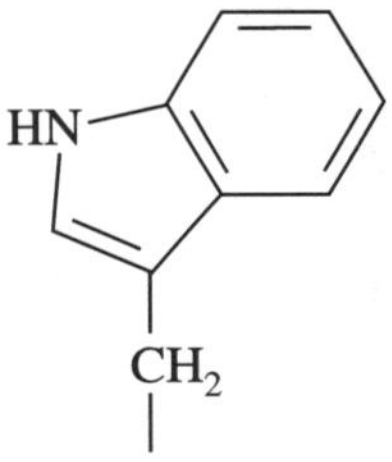 indol-3-yl-CH_2-
tyrosine	Tyr	$HO-C_6H_4-CH_2-$
valine	Val	$(H_3C)_2CH-$

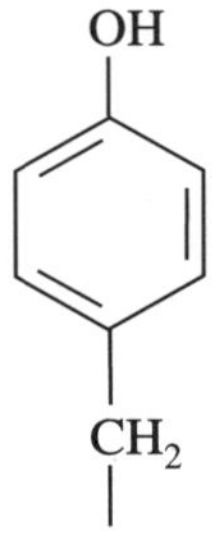

acid dissociation constant, K_a The equilibrium constant for a weak acid; a measure of the acidity of compounds; the smaller or more negative the pK_a, the stronger the acid. For a reaction

$$HX(aq) + H_2O(l) = X^-(aq) + H_3O^+(aq)$$

$$K_a = K[H_2O] = \frac{[H_3O^+][X^-]}{[HX]}$$

and $pK_a = -\log_{10} K_a$

active site The location in an enzyme where the substrate binds and is transformed to the product by the enzyme catalyst.

ATP (adenosine 5-triphosphate) ATP is an important molecule in the biological energy cycle. ATP reacts with water to give an inorganic phosphate (P_i), ADP (adenosine diphosphate) and energy (about $30 \, kJ \, mol^{-1}$). This chemical energy is converted into other forms, such as the kinetic energy of muscles, and ATP is therefore often referred to as an energy transducer.

anaemia A medical condition that is characterised by a low count of red blood cells; it often leads to tiredness and lethargy.

base (biological) (see *nucleobases*)

base sequences (see *DNA*)

biomineralisation The biological synthesis of inorganic phases. Examples are bone and shell.

carbonic anhydrase A zinc-containing enzyme that catalyses the reversible hydration of carbon dioxide to hydrogen carbonate ions and hydrogen ions.

catabolism The breaking-down of a molecule into smaller molecules, usually involving the production of ATP. Catabolism may be aerobic or anaerobic. There are many different aerobic catabolic pathways, but these all lead into the central pathway. Catabolism and anabolism together constitute metabolism.

catalase A *haemprotein* that catalyses the decomposition of hydrogen peroxide to water and oxygen.

chlorophyll A magnesium-containing *porphyrin*, often found as a pigment in photosynthetic systems in plants and green algae.

***cis*-platin** *cis*-Diamminedichloroplatinum(II), which is used in the treatment of cancers.

cristae *Mitochondria* contain a heavily folded internal membrane. The folds form pockets known as cristae. These pockets are the important sites for O_2 reaction in respiration. O_2 is delivered into the cristae, where it is bound and reduced by *cytochrome c oxidase*, which is found embedded in the membrane.

cytochrome c oxidase (CCO) An iron-containing enzyme found in the mitochondrial *cristae*, where the enzyme reduces oxygen to water. The reaction is the final step of aerobic respiration.

cytochrome P450 A *haemprotein* that catalyses the mono-oxygenation of organic molecules. It is called P450 because the carbonylated protein has a strong absorption in the visible spectrum at 450 nm.

cytochromes Iron-containing proteins, which are the electron carriers in the electron transport chain. The iron atom within each cytochrome molecule is alternately oxidised to iron(III) and reduced to iron(II). The cytochrome at the end of the chain (cytochrome c oxidase) is directly involved in the reduction of oxygen to water.

diffusion The movement of a solute through a medium or through a barrier such as a membrane. The solute moves from a region of higher concentration to one of lower concentration. The solute may be a gas, ions, etc.

DNA (deoxyribonucleic acid) A hetero-polymer in which the monomer is a composite molecule consisting of a phosphate group joined to a deoxyribose molecule, which in turn is joined to one of four different purine/pyrimidine bases (*nucleobases*) — adenine, guanine, cytosine or thymine. The monomers are called deoxyribonucleotides. The DNA molecule has a very characteristic double helix structure, made from two polymer strands intertwining around each other. In *eukaryotic* cells, the DNA of the nucleus is found, in conjunction with histones, in chromosomes. Such DNA carries the genetic information of the cell, coded as a sequence of deoxyribonucleotides (and hence as a sequence of bases).

dioxygenase An enzyme that inserts *both* atoms of O_2 (one of its substrates) into another substrate (usually organic).

double helix See *DNA*.

enterobactin A *siderophore* secreted by bacteria for the chelation of iron.

enzyme A macromolecule that catalyses a reaction. More precisely, it is a biological catalyst composed entirely or partially of protein. A given enzyme is able to catalyse a particular reaction or type of reaction. The names of enzymes end in -ase, the preceding part of the name usually being the name of the substrate or the class of substrate or the type of reaction catalysed, or some mixture of these. Enzymes are coded for by structural genes.

Escherichia coli A *prokaryotic* bacterium; often abbreviated to *E. coli*.

eukaryote An organism whose cells contain a nucleus, within which lie the chromosomes.

EXAFS (extended X-ray absorption fine structure) An X-ray absorption technique that is capable of determining the local radial structure of a particular element. Can be used with non-crystalline samples.

ferritin An iron-storage protein found in mammals. Iron is stored mostly as hydrated iron(III) oxide in the protein.

haem The non-polypeptide part of *haemoglobin* and *myoglobin*. An iron–*porphyrin* complex involved in oxygen transport.

haemerythrin An O_2-transport protein found in certain marine invertebrates.

haemochromatosis A medical condition of massive iron overload in the body.

haemocyanin An O_2-transport protein found in some molluscs and arthropods such as snails and spiders, respectively.

haemoglobin An O_2-transport protein found in the red blood cells of mammals. Each haemoglobin molecule is composed of four polypeptide chains, each with an oxygen-carrying haem group.

haemosiderin An iron-storage protein found in some mammals. It stores iron mostly as hydrated iron(III) oxide. The three-dimensional structure of haemosiderin appears to be variable and complex.

haemprotein A protein that contains one or more *haem* groups.

hard/soft acid–base theory A theory developed by R. G. Pearson, which states that stable complexes are formed between hard acids and hard bases, and between soft acids and soft bases. Hard acids and bases are characterised by small size and high absolute charge. Soft acids and bases are more polarisable, and are characterised by large size and low absolute charge.

higher-order structure The overall three-dimensional structure of a biochemical molecule, divided into secondary, tertiary and quaternary structures. The shape is unique for each protein and arises as a consequence of a particular arrangement of weak bonds between residues at various places in the primary structure. Thus, the primary structure of a protein (its sequence of amino acid residues) strongly influences the higher-order structure. Secondary structures, such as the α-helix and β-sheet, are stabilised by interactions of the backbone atoms rather than the side-chains. The weak bonds stabilising secondary structures are often hydrogen bonds. The tertiary structure refers to the extensive folding of the primary structure, which usually results in a compact molecule. Tertiary structure of a protein, for example, consists of a mixture of secondary structure and irregularly folded regions. Quaternary structure is the three-dimensional arrangement of different polypeptide chains. The double helix of *DNA* is another example of regular secondary structure, stabilised by hydrogen bonds between base pairs and hydrophobic interactions between stacked bases.

hormone   A chemical messenger produced in very small quantities in one part of an organism and transported to the target tissue where it exerts its effect.

lactoferrin   An iron-transport protein found in human breast milk.

Lewis acid   A molecule or ion that is able to accept a pair of electrons.

Lewis base   A molecule or ion that is able to donate a pair of electrons.

metalloprotein   A protein that contains one or more metal atoms, which are essential for its activity.

mitochondrion (plural mitochondria)   *Eukaryotic* cell component that is a centre of respiration; small subcellular organelle concerned with oxidative catabolism. They contain, among others, the enzymes of the link reaction, the tricarboxylic acid cycle, the β-oxidation pathway and the electron transport chain. As they are the principal ATP-making organelles, mitochondria are often termed the powerhouses of the cell.

myoglobin   An O_2-storage protein found in muscle tissue. Each myoglobin molecule consists of a single polypeptide chain to which a *haem porphyrin* group is attached.

NAD$^+$ Nicotinamide adenine dinucleotide. A coenzyme that acts as a hydrogen acceptor in many enzyme reactions that involve dehydrogenation.

nucleobase (in *nucleotides*)   A general name, used in the context of nucleic acids, for those molecules that have purine and pyrimidine ring structures. Adenine and guanine are purine bases; thymine, cytosine and uracil are pyrimidine bases.

adenine

guanine

cytosine

uracil
(RNA only)

thymine
(DNA only)

nucleotides   Monomers that, condensed together, form a nucleic acid. Each molecule contains a purine or pyrimidine base bound to a sugar group (usually a ribose or 2-deoxyribose), which is, in turn, bonded to one or more phosphate groups. The term is often loosely used to mean the nucleotides found in either DNA (deoxyribonucleotides) or in RNA (ribonucleotides).

peroxidase   A *haemprotein* that catalyses the oxidation of a substrate by hydrogen peroxide.

picket-fence porphyrin   A synthetic haem group that has a fence-like structure to prevent dimerisation of the haem groups once oxygenated.

polypeptide   A polymer formed from a selection of the 20 or so naturally occurring α-amino acids. The term 'polypeptide' is used instead of protein (a) when referring to a newly synthesised linear chain or (b) when referring to one of the component chains of a protein containing two or more polypeptide chains per molecule e.g. haemoglobin. Sometimes polypeptide is used interchangeably with protein.

porphyrin   A cyclic molecule made from four pyrrole (or pyrrole-like) rings. Porphyrins are often found in biochemical systems with a metal ion coordinated at their centre.

prokaryote   An organism whose cells do not contain a nucleus. Prokaryotes are mostly unicellular organisms, which contain ribosomes but have no well-defined subcellular structures.

protein   A large polymer of amino acids, consisting of one or more polypeptide chains. *Myoglobin* has one chain and *haemoglobin* four chains.

protoporphyrin IX   A particular *porphyrin* group that is often found in proteins. It occurs as its iron complex in *haemoglobin* and *myoglobin*.

resonance Raman spectroscopy   A variant of Raman spectroscopy. The technique gives information about the molecular vibrations associated with a chromophore (that is, the part of a molecule that gives rise to its colour) by tuning the exciting laser frequency to an electronic transition. It is particularly useful for studying transition metals in proteins.

saccharides   Molecules containing carbon, hydrogen and oxygen with a large number of hydroxyl groups in addition to a carbonyl group. The monomers have the molecular formulae $(CH_2O)_n$, where n is commonly 5 or 6. Examples are glucose, fructose and ribose. They condense to form polymers known as polysaccharides. Collectively, saccharides are also known as carbohydrates.

siderophore   A molecule that is secreted by bacteria for the chelation of iron.

stability constant   A value representing the equilibrium constant of the reaction of a ligand with a metal ion. The value of the stability constant is a measure of the stability of the metal–ligand complex.

substrate   Usually, the term substrate is used to describe a chemical species that interacts specifically with the active site of an enzyme, where it undergoes chemical transformation.

superoxide   The O_2^- anion, which contains an unpaired electron and can, therefore, be classed as a radical. It is very reactive.

superoxide dismutase   A class of enzymes that catalyses the decomposition of the superoxide anion, O_2^-, to oxygen, O_2, and hydrogen peroxide, H_2O_2.

thalassaemia   A genetic disease with anaemia-like symptoms. The disease is caused by a defective haemoglobin structure. Some types of thalassaemia can be severe.

transferrin   An iron-transport protein.

vitamin   An organic compound essential (in small quantities) for the health of an organism. Consequently, supplies have to be obtained in the diet.

X-ray diffraction   A technique that can be used to determine the accurate molecular structure of compounds. Single crystals of the compounds are needed for this technique.

zinc-finger protein   A protein that contains a finger-like loop structure; this structure is essential for the protein's function, which is to bind to DNA. Zinc is an essential structural element of the protein.

OBJECTIVES FOR BLOCK 7

Now that you have competed Block 7, you should be able to do the following things:

1 Recognise valid definitions of, and use in a correct context, the terms, concepts and principles in Table A.

Table A List of scientific terms, concepts and principles used in Block 7

Term	Page No.
absorption edge	25
allosteric interaction	44
biomineralisation	17
bulk element	6
carbonic anhydrase (CA)	80
catalases	60
chelate effect	66
collagen	16
cooperative effect	44
coordination shells	27
cytochrome c oxidase (CCO)	55
diffusion	30
distal histidine	39
electron transport	20
end-on bridging	34
end-on coordination	34
entatic state	20
esterases	73
extended X-ray absorption fine structure (EXAFS)	25
ferritin	76
haem group	38
haemerythrin (Hr)	48
haemochromatosis	63
haemocyanin (Hc)	48
haemoglobin (Hb)	35
haemosiderin	76
homeostasis	64
iron picket-fence porphyrin	47
iron overload	63
isomorphous replacement	37
lactoferrin	75
ligand preorganisation	72
liver alcohol dehydrogenase (LADH)	82
macrocyclic effect	67
mono-oxygenases	58
myoglobin (Mb)	36
passive diffusion	30
peroxidases	60
Perutz trigger mechanism	45
porphyrin group	21
positive cooperative effect	44

Term	Page No.
proximal histidine	39
R-state	45
radial distribution curve	26
recommended daily allowances (RDA)	6
relaxed form	45
resonance Raman effect	28
resonance Raman spectroscopy (RR)	28
respiratory burst	57
side-on bridging	35
side-on coordination	33
siderophore	68
single-crystal X-ray diffraction (XRD)	24
solubility product	7
spectroscopically silent compound	79
synchrotron	25
synchrotron radiation	25
T-state	45
tense form	45
tetrapyrroles	21
thalassaemia	62
trace element	6
transferrin	74
X-ray absorption spectrum	25
zinc-finger proteins	84

2 Discuss why inorganic elements are an essential part of biochemistry. (SAQs 31 and 39)

3 Show, with examples, that metals in biological molecules are often coordinated by certain amino acid side-chains, e.g. histidyl, aspartyl, glutamyl, cysteinyl, methionyl, tyrosyl. (SAQs 2 and 3)

4 Show that many biological molecules (e.g. haemoglobin) contain inorganic elements, and that these molecules have chemistry that is related to simple complexes of the inorganic element. (SAQs 4, 7, 8, 31, 35 and 36)

5 Show, with examples, that biological molecules may also contain other non-amino acid ligands, such as porphyrins, catechols, sugars and nucleobases. (SAQ 4)

6 Describe some experimental techniques, that are used to study metalloproteins. (SAQs 6, 7, 8, 10, 11, 15 and 29)

7 Describe some of the proteins involved in O_2 transport and activation. (SAQs 1, 9, 12, 13, 14, 15, 16, 17, 18, 19, 20 and 33)

8 Show that iron is an essential part of many enzymes and proteins, and that the protein's function is often dependent on a change in oxidation state of the iron. (SAQs 1, 9, 14, 16 and 30)

9 Describe some of the sophisticated biochemical methods for the uptake, transport and storage of iron, including microbial siderophores, and mammalian transferrin and ferritin. (SAQs 21, 22, 23, 24, 25, 32, 37 and 38)

10 Show, with examples, that zinc is an important bioinorganic element with distinct roles, different from those of iron. (SAQs 1, 26, 27, 28 and 30)

11 Discuss the future of bioinorganic chemistry, especially the use of inorganic compounds as drugs/medicines.

SAQ ANSWERS AND COMMENTS

SAQ 1
(Objectives 7, 8 and 10)

Protein	Metal(s)	Relative molecular mass	Function
light harvesting complex	Mg		photosynthesis protein
haemoglobin	Fe	64 500	O_2 transport
myoglobin	Fe	17 800	O_2 storage
haemerythrin	Fe	13 500	O_2 transport
haemocyanin	Cu	1 000 000	O_2 transport
cytochrome c oxidase	Fe/Cu	100 000	catalyses reduction of O_2 to H_2O
cytochrome c	Fe	12 000	transfers electrons to CCO
cytochrome P450	Fe		O_2 activation/mono-oxygenase
superoxide dismutase	Cu/Zn		catalyses $O_2^{-}\cdot \longrightarrow O_2 + H_2O_2$
catalase	Fe		$O_2^{-}\cdot$ disproportionation
peroxidase	Fe		mono-oxygenase using H_2O_2
siderophores	Fe		iron transport
transferrin	Fe	80 000	iron transport
ferritin	Fe	440 000	iron storage
haemosiderin	Fe		iron storage
lactoferrin	Fe		iron transport
carbonic anhydrase	Zn	30 000	CO_2 hydration
liver alcohol dehydrogenase	Zn		catalyses alcohol oxidation to aldehydes
zinc-finger protein	Zn		DNA binding

SAQ 2
(Objective 3)

Serine (side-chain —CH_2OH) and threonine (side-chain —$CH(OH)CH_3$) are possibilities in that they may coordinate via their oxygen lone pairs.

Glutamine (side-chain —$CH_2CH_2C(O)NH_2$) and asparagine (side-chain —$CH_2C(O)NH_2$) are possibilities, although coordination will occur through the lone pair(s) on the oxygen atom, not the nitrogen atom. (The reason for this is the same reason why the amide group of the polypeptide chain is a poor ligand.)

Lysine (side-chain —$CH_2CH_2CH_2CH_2$–NH_2) is also a possibility.

Arginine (side-chain —$(CH_2)_3NHC(=NH_2^+)$—NH_2) is positively charged at pH 7, so does not form stable complexes with positive metal ions.

Tryptophan formally has a nitrogen lone pair of electrons. However, this lone pair forms part of a π molecular orbital in the side-chain (similar to the amide bond), and is unavailable for coordination to a metal ion. (You may also have said that the benzene ring of tryptophan could coordinate to a metal in η^6 fashion as in the metallocenes you met in Block 6, but this would be very unlikely in an aqueous medium.)

It is also possible that the terminal amino and carboxylate groups of a polypeptide chain could coordinate to metals via the –NH_2 or CO_2^- group, respectively.

SAQ 3
(Objective 3)

As copper(I) is a soft metal ion we would expect that it would form stable complexes with soft ligands. In fact, in biochemical systems, several proteins involved in electron transfer contain copper(I) ions at the active site. In these proteins, the copper is coordinated by the sulphur atom of at least one methionyl or cysteinyl side-chain.

As copper(II) is in a higher oxidation state, it is a harder metal ion than copper(I), and ligands that stabilise the copper(II) ion are also expected to be hard. Such ligands include the amino acid residues aspartyl, glutamyl and histidyl.

SAQ 4
(Objectives 4 and 5)

Pt^{2+} is a soft metal. If it were to interact with DNA, it would be effectively coordinated by nitrogen atoms of a nucleobase. This would lead to a disruption of the regular DNA helical structure. Structurally damaged DNA is potentially carcinogenic (see Section 9).

SAQ 5

Al^{3+} has the high charge-to-radius ratio of 0.044 pm^{-1}. The metal with the nearest value to this is Fe^{3+} (0.038 pm^{-1}). If we realise that the charge-to-radius ratio reflects the stability of a metal ion with a given ligand (similar to the Irving–Williams series), then we can see that if Al^{3+} is taken up by the body, it will be coordinated by the same biochemical ligands that chelate iron. This is, indeed, thought to be the case, where aluminium is incorporated in the body's iron uptake and transport systems (see Section 7).

Aluminium is essentially redox inactive; in other words it does not form stable oxidation states other than +3. (*Note* Al(O) occurs on the surface of metallic aluminium, but in the context of bioinorganic chemistry this form of aluminium is not important.) In contrast, iron is redox active, with its +2 and +3 oxidation states being the most common. Indeed, iron is prevalent in biochemistry partly because of its redox activity. Replacement of iron by aluminium in biochemistry, eventually affords proteins that do not display their normal redox activity. As biochemical processes are so finely balanced, any changes to the normal activity of proteins can lead to dramatic biochemical effects. It is clear from this that aluminium poisoning can have severe biochemical consequences, since iron is found in very many key proteins in biochemistry (as you will find throughout this Block).

SAQ 6
(Objective 6)

Vibrations of ligands *directly attached* to the metal ion give rise to Stokes and anti-Stokes lines in the scattered light. In this way we can selectively study the metal ion and the *molecular vibrations* of its immediate environment within a protein. As such, the technique is able to concentrate on the metal site in the protein. The spectrum obtained from RR studies of proteins can usually be assigned easily.

SAQ 7
(Objectives 4 and 6)

Simply because these techniques give information about the complete protein. The spectra obtained have so many overlapping bands that they cannot be interpreted.

SAQ 8
(Objectives 4 and 6)

The advantage of XRD is that it gives a full structural picture of the protein under study. Without knowledge of a protein's structure, it is difficult to understand how the protein functions. One disadvantage of this technique is that crystals of the protein need to be grown: this may be very difficult, if not impossible. Another disadvantage is that the technique requires expensive equipment and a large time investment by a range of specialists.

SAQ 9
(Objectives 7 and 8)

When free haem is reacted with O_2, the first product of the reaction is a molecule that has two haem groups bridged by a μ-η^1,η^1 oxygen molecule. A similar molecule is shown as the product of reaction 55:

SAQ 10
(Objective 6)

If the O_2 were indeed bound as superoxide (that is, O_2^-), which it must be if the iron has been oxidised to iron(III), then we would expect the vibrational frequency to be less than that of O_2, because the electron would go into the $2p\pi_g$ antibonding orbital, thus weakening the O—O bond. In fact, from resonance Raman studies on real myoglobin, the vibrational frequency of the O_2 in oxy-Mb has been measured at about $1\,110\,cm^{-1}$, where we would expect a superoxide vibration frequency to appear (see Figure 26).

SAQ 11
(Objective 6)

(a) We could not use infrared spectroscopy to study the vibrations of O_2 because its dipole moment does not change during vibration. Raman spectroscopy can be used instead.

(b) We would expect the vibrational (Raman) spectrum to contain three bands in the ratio 1 : 2 : 1. The O_2 molecules with the heavier ^{18}O isotope will vibrate at a lower frequency than the $^{16}O-^{16}O$ molecule; remember that the vibrational frequency is inversely proportional to the reduced mass

$$\nu = \frac{1}{2\pi}\sqrt{\frac{k}{\mu}}$$

56

Figure 85 shows the Raman spectrum of 50 per cent ^{18}O-labelled O_2.

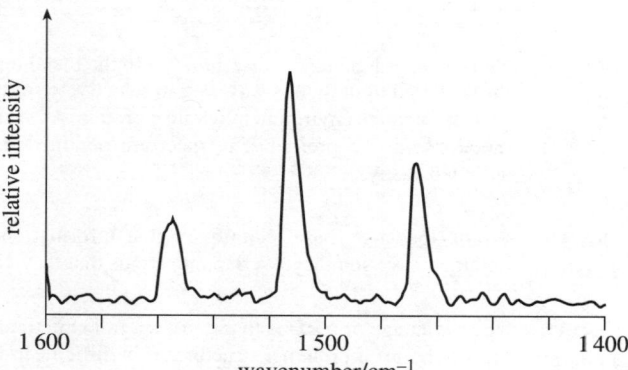

Figure 85 Raman spectrum of 50 per cent ^{18}O-labelled O_2 gas.

(c) The band at just below $1\,560\,cm^{-1}$ corresponds to an $^{16}O-^{16}O$ vibration. The large central band at about $1\,510\,cm^{-1}$ is due to an $^{18}O-^{16}O$ vibration and the last, lowest-frequency band, at $1\,470\,cm^{-1}$ corresponds to a $^{18}O-^{18}O$ vibration.

SAQ 12
(Objective 7)

$820\,cm^{-1}$ is the wavenumber characteristic of the peroxide ion, O_2^{2-} (Figure 26), so the charge on the bound O_2 will be 2–. It appears as if the two copper atoms have each been formally oxidised to copper(II) in the binding process.

SAQ 13
(Objective 7)

There are two properties that a transition metal complex should have in order to bind oxygen reversibly. These are a vacant coordination site on the metal to which O_2 can bind, and the ability to change oxidation state. The first is dependent on the nature of the ligand. So, for example, a transition metal–edta complex is very unlikely to bind oxygen, simply because no free coordination sites are available on the metal for the initial binding of O_2. The second property, a change in oxidation state, is, of course, a characteristic feature of transition metals.

So, in principle, any transition metal complex whose metal can change oxidation state and bind an oxygen molecule is capable of reversible oxygen binding. In reality, however, it is not so simple; much depends on the ligands and the redox potential of the metal complex. It turns out that, besides iron and copper, the transition metal complexes that are most likely to show reversible oxygen binding contain cobalt, vanadium or iridium.

SAQ 14
(Objectives 7 and 8)

This question is answered by looking back at Figure 39. The Figure shows the differences in binding geometries of CO and O_2 to the haem group in Mb. *In the absence of any other protein structure*, the most stable geometry of the iron–CO group is linear; in the absence of the distal histidine this would be a very stable complex. However, the rigid, steric bulk of the histidyl means that it is impossible for the iron–CO group to have a linear geometry in Mb. Therefore, the CO complex is *destabilised*. Furthermore, the distal histidine is able to form a hydrogen bond with any oxygen that binds to the iron, thus *stabilising* the O_2 complex.

Changing the histidine to tyrosine, changes the steric bulk only slightly, so the linear iron–CO complex would still be destabilised. Also, tyrosine would be able to form a hydrogen bond via its hydroxyl group to O_2 in the oxygenated Mb. Consequently, changing the distal histidine for tyrosine would be expected to have little effect.

Valine, on the other hand, has no hydrogen-bonding potential, and does not have as much steric bulk as either histidine or tyrosine. Therefore, we would expect that replacing the distal histidine with valine would result in a Mb that had a relatively high affinity for CO. This has been demonstrated experimentally.

SAQ 15
(Objective 7)

To determine the mode of O_2 binding in the dimer we could perform a Raman experiment. It is unlikely that a resonance Raman experiment would be needed, since the molecule is simple enough to assign the whole vibrational spectrum. In other words, we could easily determine which Raman absorption corresponded to the O—O vibration frequency. The Raman experiment would tell us the frequency of the O—O vibration, which has a direct correlation with the formal charge on the O_2. (See Figures 25 and 26 for more information.) This information would also give us an indirect indication of the oxidation state of the cobalt atoms. For example, if the Raman experiment showed an O_2 vibration in the range 720–900 cm^{-1}, we could conclude that the O_2 was bound as a peroxide (O_2^{2-}). Necessarily, this would mean that the cobalt atoms would have been oxidised. Probably both cobalts would be cobalt(III), although we cannot be certain.

Determining the O_2 binding geometry is more difficult. A Raman experiment on the oxygenated complex prepared with 50 per cent ^{18}O-labelled O_2 would help to determine whether the O_2 was bound symmetrically or unsymmetrically between the two cobalts. However, to determine the geometry unambiguously we would need to perform a single-crystal X-ray diffraction experiment on crystals of the oxygenated complex. This experiment would determine the relative positions of each atom in the oxygenated complex.

SAQ 16 *(Objectives 7 and 8)* From the answer to SAQ 14, it is clear that the binding of CO to Mb (or Hb) is destabilised. However, the affinity of Mb for CO is still 250 times *greater* than for O_2. (In the absence of the distal histidine, Mb's affinity for CO is 25 000 times greater than for O_2.) The reason that our Mb is not completely saturated with CO is that the usual atmospheric concentration of CO is much less than that of O_2. This very large preponderance of O_2 ensures that our Mb is more likely to bind O_2 from air than CO. Any CO that is present in the air we breathe will be bound, but as it is in such low quantities it makes very little difference. It has been estimated that under 1 per cent of our total Mb and Hb has bound CO.

However, there is a clear danger here. Any increase in the concentration of CO in air can lead to a much higher fraction of Mb and Hb with bound CO. This, of course, will reduce the oxygen supply to our bodies and may eventually lead to death. Higher concentrations of CO are found near busy roads (CO is emitted from car exhausts, especially from cars that do not have a catalytic converter) and in rooms with poorly ventilated gas fires. These high concentrations of CO can be lethal and must be carefully avoided. (And of course, smokers constantly poison themselves because of the higher concentration of CO in cigarette smoke.)

SAQ 17 *(Objective 7)* The intermediate is likely to be a porphyrin–iron(V)–oxo species (structure **43**), similar to those seen in CCO and P450. This intermediate reacts rapidly with H_2O_2 in the second step to give O_2 and H_2O.

43

SAQ 18 *(Objective 7)* The mechanism is very similar to that for catalase, and can be broken down into two steps. Again, the key enzyme intermediate is a porphyrin–iron(V)–oxo species.

peroxidase–iron(III) + H_2O_2 = peroxidase–iron(V)–oxo + H_2O 57

peroxidase–iron(V)–oxo + XH_2 = peroxidase–iron(III) + H_2O + X 58

SAQ 19 *(Objective 7)* Both of these enzymes must react with the O_2-containing molecule before it can react with any other molecule. Therefore, the enzymes must be very efficient, catalysing the decomposition of every substrate molecule that they encounter. In other words, they must operate near to the diffusion limit.

SAQ 20 *(Objective 7)* To answer this question we need to turn back to Figure 59. Firstly, we must replace the substrate, RH in Figure 59, with the camphor molecule. Doing this brings us to the situation where we have a camphor molecule adjacent to the highly reactive iron(V)–oxo species (structure **26**). At this point the oxygen atom *inserts* into a camphor C—H bond. However, it only inserts into the C—H bond in the 5-position of the camphor molecule. Moreover, only the *exo*-product is obtained. The

SLC 7 reason for this high *regio- and stereoselectivity* is that the active site orientates the camphor molecule into the correct position before reaction. In other words, the camphor molecule is held in the active site with the *exo*-C—H bond nearest the iron(V)–oxo species.

SAQ 21
(Objective 9)
There are three key points. Firstly, enterobactin is a hexadentate ligand and so the stability of the complex is enhanced by the chelate effect. Secondly, the groups that coordinate to iron(III) are hard ligands. Thirdly, the enterobactin ligand is preorganised and must be exactly the correct size, shape and charge for iron(III) binding.

SAQ 22
(Objective 9)
A siderophore is a microbial iron chelator, whose complex with iron(III) usually has a very high stability constant. Examples include enterobactin, aerobactin and agrobactin.

SAQ 23
(Objective 9)
As with enterobactin, both **38** and **39** will coordinate to iron(III) via the six deprotonated oxygens in each structure (highlighted in green below).

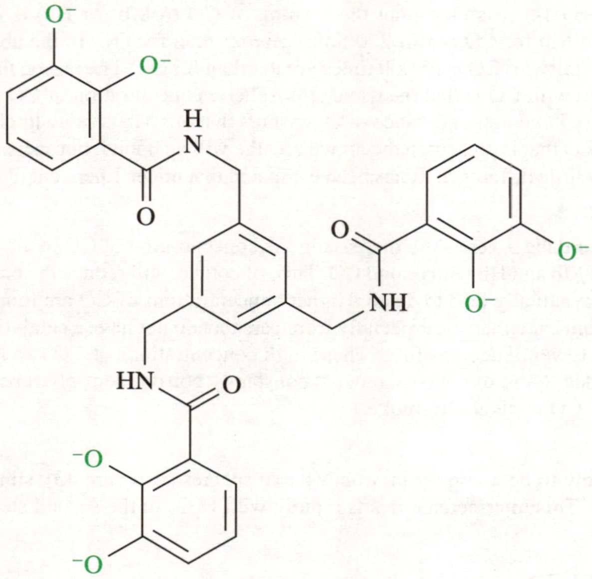

38 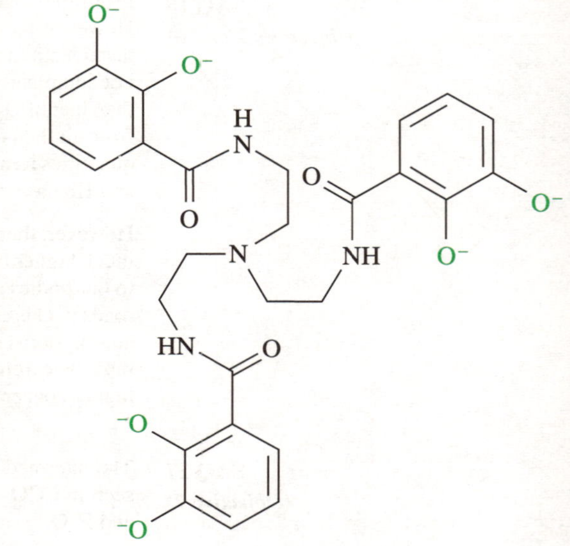 **39**

The key feature of **38** and **39** that distinguishes them from enterobactin is the replacement of the triserine ring with 1,3,5-trimethylbenzene in **38** and a triethylamine in **39**. The effects are twofold. Firstly, both **38** and **39** are not as rigid (preorganised) as enterobactin. Secondly, both **38** and **39** have different relative spacing of their catechol groups compared to enterobactin. In fact, on chelation of iron(III), the catechol groups in **38** and **39** cannot obtain the most stable spatial arrangement around the iron, because their motion is restricted by the 1,3,5-trimethylbenzene in **38** and the triethylamine in **39**. Hence the stability constants of the complexes formed by **38** and **39** with iron(III) are lower than for enterobactin. **38** has a slightly higher stability constant with iron(III) than **39** for two reasons. Firstly, **38** is slightly more rigid than **39** and not so much ligand reorganisation is required on complex formation. Secondly, the size of the trimethylbenzene is slightly larger than that of the triethylamine, and the catechol groups in **38** are able to approach the most stable spatial arrangement around the iron(III) ion better than in **39**.

SAQ 24
(Objective 9)
Structure **44** shows the expected conformation of agrobactin in its iron(III) complex, where the iron is chelated by the three deprotonated catechol groups, in an analogous fashion to the iron(III) binding with enterobactin.

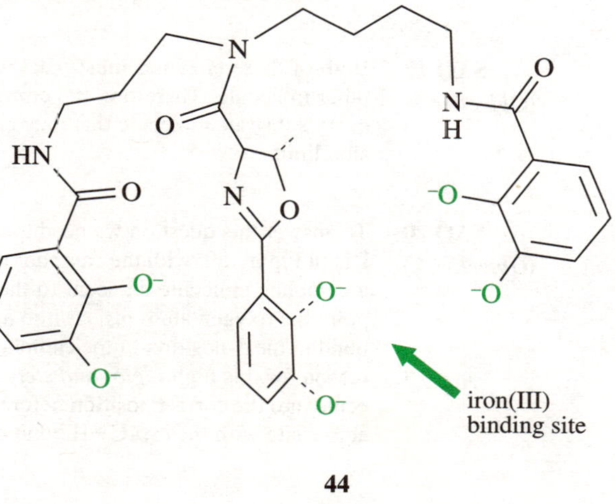

44

SAQ 25
(Objective 10)

No other element could replace zinc in proteins without some reduction in the proteins' catalytic ability. We have already seen the successful substitution of zinc(II) by cobalt(II) in zinc-finger proteins. Looking at the other first-row transition metals, nickel(II) has a stable +2 oxidation state and will form tetrahedral complexes (although with appropriate ligands, nickel(II) often forms square-planar complexes). Other first-row transition metals with a stable +2 oxidation state — manganese(II), iron(II) and chromium(II) — are too redox active to be of use. If we can ignore the +2 charge, then another possibility is copper(I) (although this would have a much lowered Lewis acidity), which readily forms tetrahedral complexes. Another possibility is cadmium(II). (Palladium(II) and platinum(II) form mostly square-planar complexes.) Silver(I) and gold(I) also form tetrahedral complexes.

SAQ 26
(Objective 10)

Recalling the catalytic cycle of carbonic anhydrase in Figure 76, we see (structure **45**) that one of the intermediates is a metal-bound hydrogen carbonate.

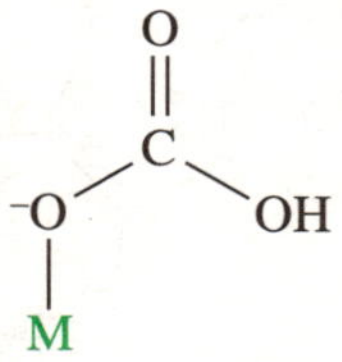

45 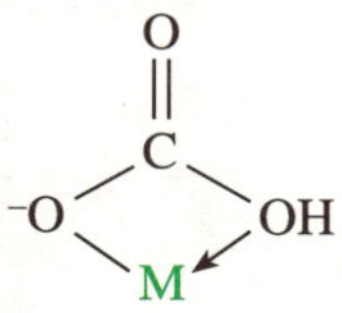   **46**

For the catalytic cycle to be completed, the hydrogen carbonate must be displaced by water. This displacement is rapid if the hydrogen carbonate is monodentate as shown in structure **45**. However, if it binds as bidentate hydrogen carbonate (structure **46**), then the displacement is much slower.

Whether the hydrogen carbonate is monodentate or bidentate depends on the nature of the metal. Copper(II) and nickel(II) will probably form bidentate complexes, because both metals readily form six-coordinate complexes. Zinc(II) will probably form a monodentate complex, since the most common zinc(II) coordination geometry is four-coordinate tetrahedral. Cobalt(II) will probably form both types, or an intermediate form — known as *anisobidentate*, since cobalt(II) will form both four-coordinate tetrahedral and five/six-coordinate complexes (Figure 86).

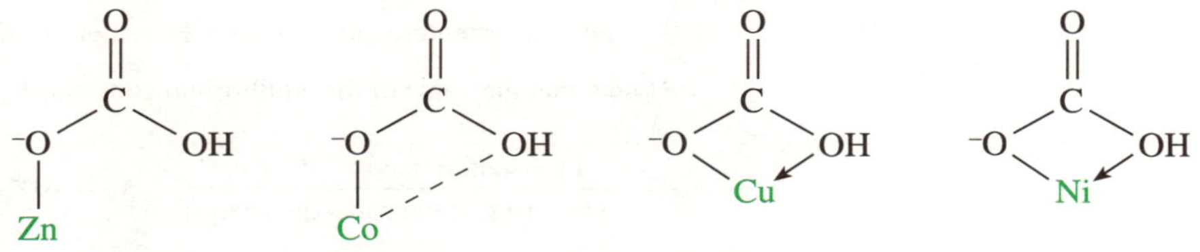

Figure 86 Variation in mode of binding a hydrogen carbonate ligand for the metals zinc(II), cobalt(II), copper(II) and nickel(II).

The tendency of the hydrogen carbonate to be coordinated in a monodentate fashion matches the observed activity of the metal-substituted carbonic anhydrase.

SAQ 27
(Objective 7)

The first point to address is what type of metal should be used to bind the oxygen directly. The metal ion must exhibit two oxidation states. As you may recall from the answer to SAQ 13, five metals fit well into this category, namely vanadium (oxidation states +2 and +3), iron (oxidation states +2 and +3), cobalt (oxidation states +2 and +3), copper (oxidation states +1 and +2) and iridium (oxidation states +1 and +3) We already know from biochemical systems that copper and iron are found in O_2-transport proteins. It is sensible therefore to pick one of these metals and to try to mimic the coordination environment found in the proteins. So, the synthetic molecule should contain either a haem group (cf. haemoglobin) or one (or two) copper(s) coordinated by three histidine-like ligands (cf. haemocyanin). The synthetic molecule also requires some type of O_2-binding site that would prevent the irreversible dimerisation of the oxygenated molecule. Therefore, as with Collman's picket-fence porphyrin, the molecule should contain some type of steric bulk around the binding site. We might also consider placing a hydrogen-bond donor in the binding site to interact with the bound O_2.

SAQ 28
(Objective 11)

It is a reasonable assumption that the biochemical ligands/proteins that coordinate iron(III) would also coordinate plutonium(IV). Therefore, we expect to find plutonium in transferrin and ferritin. If we required a ligand for the removal of plutonium from the body, we would want to design a ligand whose complex with iron(III) has a high stability constant (because, again, it is a reasonable assumption that the ligand would also form a plutonium(IV) complex with a high stability constant). We can take our clues about ligand design from the siderophores. Presumably a hexadentate chelating ligand, with hydroxamic acid and/or catechol groups, will be effective at chelating plutonium(IV). The possible side effects of treatment with this ligand are related to the fact that the ligand will also chelate a lot of iron, and may interfere with the biochemistry of iron, possibly making the patient anaemic.

SAQ 29
(Objective 6)

Figure 87a shows the plane of Fe–protoporphyrin IX with circles at 200 pm and 250 pm. We expect the EXAFS radial distribution plot to contain a large peak at about 200 pm (Figure 87b), corresponding to the four equidistant nitrogen atoms of the haem pyrrole groups (and also to the nitrogen of the proximal histidine). This peak will be close to another peak at about 250 pm, corresponding to the eight equidistant carbon atoms directly attached to the pyrrole nitrogens (and also to two carbon atoms connected directly to the proximal histidine nitrogen atom).

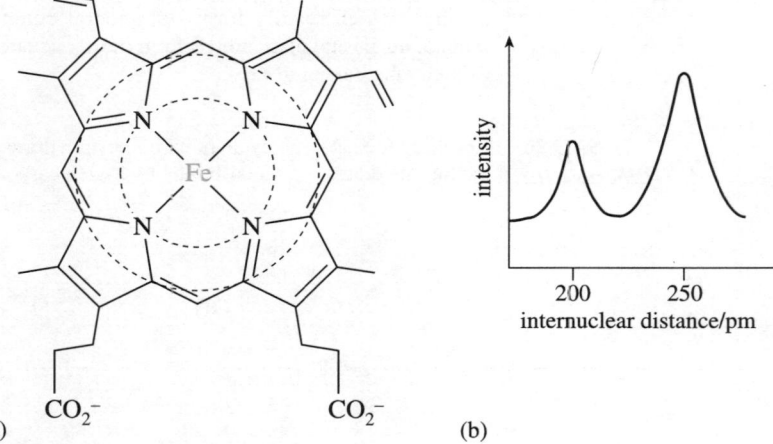

Figure 87 (a) The plane of Fe–protoporphyrin IX, with circles drawn at 200 pm and 250 pm from the nucleus of the iron atom; (b) schematic Fe-EXAFS radial distribution plot for Fe–protoporphyrin IX.

SAQ 30
(Objectives 8 and 9)

Zinc has only one stable oxidation state (other than zero), which is +2. The metal in myoglobin (iron) needs to be able to participate in single-electron transfer, which is equivalent to a change in oxidation state of +1.

SAQ 31
(Objectives 2 and 4)

Too little fluoride can lead to tooth decay and brittle bones. Too much fluoride can lead to a variety of complaints, including poor healing of wounds (Section 2).

SAQ 32
(Objective 9)

For the reaction:

$$Fe^{3+}(aq) + \text{enterobactin}^{n-}(aq) = Fe\text{–enterobactin}^{(3-n)+}(aq) \qquad 59$$

we know that the value of the equilibrium constant, $K_s = 10^{49}$ mol^{-1} l (Section 7.2), is given by:

$$\frac{[Fe\text{–enterobactin}^{(3-n)+}(aq)]}{[Fe^{3+}(aq)][\text{enterobactin}^{n-}(aq)]} = K_s = 10^{49} \text{ mol l}^{-1} \qquad 60$$

Rearranging gives:

$$\frac{[Fe\text{–enterobactin}^{(3-n)+}(aq)]}{K_s} = [Fe^{3+}(aq)][\text{enterobactin}^{n-}(aq)] \qquad 61$$

But the dissociation of the complex gives equal molar amounts of free enterobactin and free aqueous iron(III),

$$[Fe^{3+}(aq)] = [\text{enterobactin}^{n-}(aq)] \qquad 62$$

so $\quad [Fe^{3+}(aq)]^2 = \dfrac{[Fe\text{–enterobactin}^{(3-n)+}(aq)]}{K_s}$

We know that $[Fe\text{–enterobactin}^{(3-n)+}(aq)] = 1$ mol l^{-1}

so $\quad [Fe^{3+}(aq)]^2 = 10^{-49}$ mol^2 l^{-2}

and $\quad [Fe^{3+}(aq)] = 3.16 \times 10^{-25}$ mol l^{-1}

Notice that this is an incredibly low concentration! As Avogadro's number is 6.022×10^{23}, this means that there are only $(6.022 \times 10^{23} \times 3.16 \times 10^{-25}) = 0.19$ Fe^{3+}(aq) ions present in the solution(!). In other words, there is either none or one free Fe^{3+} ion in 1 litre of solution.

SAQ 33
(Objective 7)

See Figure 61 (Section 6) for the catalytic cycle. The copper in the active site acts like an electron storage site during the disproportionation of the superoxide (structures **29** and **30**). The copper's ability to form +1 and +2 oxidation states is important for this step. The zinc's role appears to be structural; it certainly has no redox activity. Its most important role is probably in maintaining the position of the bridging imidazolate group, and also ensuring that the pK_a of the imidazole group coordinated to the zinc is low enough to ensure deprotonation to imidazolate (structures **31** to **28**); in this sense, zinc is acting as a Lewis acid.

SAQ 34
(Objective 4)

As well as binding strongly to myoglobin and haemoglobin, both ligands irreversibly bind to cytochromes — especially cytochrome c oxidase — causing an immediate shut-down of aerobic respiration.

SAQ 35
(Objective 6)

Refer to Figure 38, Section 5. Here we see that there are no unpaired electrons, giving rise to a diamagnetic ground state in oxygenated myoglobin.

SAQ 36
(Objective 7)

As both copper and iron have two stable oxidation states, they can participate in one-electron transfer to O_2.

SAQ 37
(Objective 9)

Transferrin chelates the available iron(III), making it less available to microbial siderophores.

SAQ 38
(Objective 9)

We know that when iron(II) compounds are exposed to air, they are usually oxidised to iron(III). This chemical process also results in generation of the superoxide radical anion. The superoxide is very reactive and will react with organic molecules. In this case, the organic molecule is DNA, and reaction with superoxide results in cleavage of the DNA double helix. The DNA could be protected by the addition of superoxide dismutase, which would react rapidly with any superoxide, converting it to hydrogen peroxide and O_2. Unfortunately, the hydrogen peroxide generated by this enzyme could also react with iron(II) to give radical species (particularly the hydroxyl radical, HO·). Therefore, catalase would also be needed to protect the DNA fully.

SAQ 39
(Objective 2)

It was argued that the bacteria that dispose of mercury by converting it to volatile dimethylmercury can do the same with antimony and convert it to trimethylstibine ($Sb(CH_3)_3$). The trimethylstibine is similarly volatile, and may then be breathed in by a baby sleeping on the mattress. It is unclear whether this is true or not.

ACKNOWLEDGEMENTS

Grateful acknowledgement is made to the following sources for permission to reproduce material in this Block:

Figures

Figure 6: © Irving Geis, Geis Archives; *Figures 15 and 16:* courtesy of Dr G. K. Barlow, Department of Chemistry, University of York; *Figure 17:* Cramer, S. P., and Hodgson, K. O. (1979), 'X-ray absorption spectroscopy: a new structural method and its applications to bioinorganic chemistry', *Progress in Inorganic Chemistry*, **25**, John Wiley and Sons Inc., reprinted by permission of John Wiley and Sons Inc.; *Figure 18:* Ebsworth, E. A. V., Rankin, D. W. H., and Cradock, S. (1987), *Structural Methods in Inorganic Chemistry*, Blackwell Science Ltd; *Figure 19b:* reprinted with permission from Binstead, N., Cook, S. L., Evans, J., Greaves, G. N., and Price, R. J. (1987), *Journal of the American Chemical Society*, **109**, p. 3669, copyright © 1987 American Chemical Society; *Figure 22:* reprinted with permission from Clark, R. J., Stead, M. J. (1983), in Chisolm, M. H. (ed.) *Inorganic Chemistry: Toward the 21st Century*, ACS Symposium Series, **211**, p. 235, copyright © 1983 American Chemical Society; *Figure 23:* Royal Geographical Society; *Figure 28:* blood supplied courtesy of Wendy Selina, Department of Chemistry, Open University, micrography and photograph by Heather Davies, Department of Biology, Open University; *Figure 29:* courtesy of Times Newspapers Ltd; *Figure 30:* courtesy of Godfrey Argent; *Figure 33:* from *Biochemistry* 3/E by Stryer, L., © 1988 by Lubert Stryer; used with permission of W. H. Freeman and Company; *Figure 47:* reprinted with permission from Kurtz *et al.* 1976, 'Resonance Raman spectroscopy with unsymmetrically isotopic ligands', *Journal of the American Chemical Society*, **98** (16), p. 5034, copyright © 1976 American Chemical Society; *Figures 51 and 85:* reprinted with permission from Thamann, T. J., Loehr, J. S., and Loehr, T. M. (1977), 'Resonance Raman study of oxyhaemocyanin with unsymmetrically labelled oxygen', *Journal of the American Chemical Society*, **99**, p. 4188, copyright © 1977 American Chemical Society; *Figure 53:* reprinted with permission from Kitajima *et al.* 1988, 'An accurate synthetic model of oxyhaemocyanin', *Journal of the Chemical Society, Chem. Commun.*, **2**, p. 151, copyright © 1988 American Chemical Society; *Figure 62:* adapted from Friedman, M. J., and Trager, W. (1981), 'The biochemistry of resistance to malaria — How is the protective effect of the gene for sickled red blood cells explained?', *Scientific American*, **244**(3), March 1981, copyright © 1981 by Scientific American, Inc., all rights reserved; *Figure 64:* courtesy of Biophoto Associates; *Figure 72:* reprinted from *Coordination Chemistry Reviews*, **151**, St Pierre, T. G., *et al.* 'Synthesis, structure and magnetic properties of ferritin cores', p. 125, copyright © 1996, with kind permission from Elsevier Science S. A., P. O. Box 564, 1001 Lausanne, Switzerland.

Table

Table 1: Kaim, W., and Schwederski, B. (1994), *Bioinorganic Chemistry: Inorganic Chemistry in the Chemistry of Life*, John Wiley and Sons Ltd., reprinted by permission of John Wiley and Sons Ltd.

Plates

Plate 1: courtesy of Daresbury Laboratory, Cheshire; *Plate 2:* courtesy of Dr A. J. Wilkinson, Department of Chemistry, University of York; *Plate 3:* data for the structure downloaded from Brookhaven Protein Data Base, processed by Wendy Selina, Department of Chemistry, Open University, graphic generated by Mark Kesby, Design Studio, Open University.

S343 Inorganic Chemistry

Block 1   Introducing the transition elements

Block 2   Theory of metal–ligand interaction

Block 3   Transition-metal chemistry: the stabilities of oxidation states

Block 4   Structure, geometry and synthesis of transition-metal complexes

Block 5   Nuclear magnetic resonance spectroscopy

Block 6   Organometallic chemistry

Block 7   Bioinorganic chemistry

Block 8   Solid-state chemistry

Block 9   Actinide chemistry and the nuclear fuel cycle